Helion & Company Limited
Unit 8 Amherst Business Centre
Budbrooke Road
Warwick
CV34 5WE
England
Tel. 01926 499 619
Email: info@helion.co.uk
Website: www.helion.co.uk
Twitter: @helionbooks
Blog: https://helionbooks.wordpress.com/

Designed and typeset by Mach 3 Solutions (www.mach3solutions.co.uk)
Cover design Paul Hewitt, Battlefield Design (www.battlefield-design.co.uk)

ISBN 978-1-804515-60-0

British Library Cataloguing-in-Publication Data
A catalogue record for this book is available from the British Library

CONTENTS

Abbreviations and acronyms 2
Introduction 3

1 Holding the Peace 3
2 The J-20 From Conceptualisation to Serial Production 16
3 China's First Stealth Fighter in Service 37
4 The J-20's Future and its Supporting Assets 44
5 International Challenges and Responses to the J-20 59

A note on writing and researching this book, and on sources of reference used 69
Notes 70
About the Author 82

I0893461

Note: In order to simplify the use of this book, all names, locations and geographic designations are as provided in *The Times World Atlas*, or other traditionally accepted major sources of reference, as of the time of described events.

ABBREVIATIONS AND ACRONYMS

AESA	active electronically scanned array
AEW&C	airborne early warning and control
AI	artificial intelligence
BVR	beyond-visual-range
CCTV	China Central Television
EOTS	Electro-Optical Targeting System
IRST	infrared search and track
NATO	North Atlantic Treaty Organization
NGAD	Next Generation Air Dominance
PACAF	Pacific Air Forces (United States)
PESA	passive electronically scanned array
PLA	People's Liberation Army
PLAAF	People's Liberation Army Air Force
R&D	research and development
UAV	unmanned aerial vehicle
U.S.	United States
USAF	United States Air Force
USSR	Union of Soviet Socialist Republics

INTRODUCTION

In the late 1970s the United States and the Soviet Union initiated rival programs to develop the first and most capable fifth generation stealth fighters. For decades the two superpowers' top fighters and interceptors were unrivalled in air-to-air combat, from the MiG-15 and F-86 of the Korean War-era to the Su-27 and F-15 of the late Cold War years, with the tremendous advantages they had over all third parties in the scales of their industries, tech sectors and defence budgets leaving little doubt that this would continue in perpetuity. The USSR's disintegration in 1991, however, was followed by both a near total collapse of Russian stealth fighter development efforts and major cuts and delays to American programs, providing an opening for China's fast modernising defence sector to strengthen its own standing in the field and develop a leading next-generation fighter of its own.

China's emergence as the world's largest industrial power in 2010, largest economy in 2014, leading patent filer in 2020, and the greatest defence acquisitions spender that year, facilitated a fast-improving standing in the large majority of strategic technologies and industries. The country has thus been increasingly well placed to lead the world not only in the capabilities of its top-end weapons systems, but also in how quickly and smoothly it can bring them into service. The J-20 fifth generation fighter has accordingly consistently far exceeded expectations in both its development timeline and its performance, requiring well under half the time of its American competitors between its first flight and service entry, while showing signs not only of parity with its top overseas competitors, but increasingly of distinct capability advantages. The fighter program thus represents a major landmark not only in the history of Chinese aviation, but also more broadly in the evolving balance of power between East and West due to the prevailing trends in industry, economy and high-tech which have facilitated its realisation.

The J-20 today is the world's second most numerous fighter of its generation, outnumbered only by the much smaller American F-35, with the two rivals seeing their first of many likely encounters confirmed in March 2022. As the J-20's capabilities have rapidly improved and the rate of production has significantly grown, it has very conspicuously played a growing role in patrolling major hotspots in East Asia from Korea and the East China Sea to the Taiwan Strait and disputed South China Sea. The fighter has been the most iconic symbol of China's ascent from an effective non-competitor at the cutting-edge of defence technologies during the Cold War, to the United States' sole peer-level competitor today. With the two defence sectors now competing neck-to-neck to field the world's first and most capable sixth generation fighter from around 2030, comparisons of the J-20's development and performance with those of the rival F-35 provide important indicators of the respective advantages both countries are expected to maintain.

This book traces China's rationale for developing a top-end air superiority fighter back to the 1950s, when almost immediately after the foundation of the People's Republic in 1949 the new state's air force faced its trial by fire in the Korean War. It compares the J-20's performance with rival fifth generation fighter programs, details how it has and will likely be fielded in the future, and explores many of its unique design features and capabilities and the rationales for their inclusion – from its particularly wide combat radius to its innovative missile launch system and adoption of a radical delta-canard airframe design. The book also details how the United States military in particular has responded to the J-20's emergence with decisions affecting training and acquisitions, explores a range of other Chinese assets developed in parallel to the new fighter and their key contributions to operational effectiveness, and looks ahead to what the future may hold for the emerging Sino-American arms race in the air.

1
CHINESE SECURITY AND THE NEED FOR AIR SUPERIORITY FIGHTERS

AIR WAR OVER KOREA

Forging Chinese combat aviation into a world power at the dawn of the jet age, forming the core of its historical memory, and shaping its strong emphasis on air superiority, the legacy of the Korean War has continued to profoundly influence China's twenty-first century military modernisation efforts including its first stealth fighter program. For the armed forces of the People's Republic of China, the People's Liberation Army (PLA), the conflict represents to date its only protracted war against state adversaries, with the pivotal role played by air power and importance of defensive air superiority in that era continuing to influence Chinese thought on air power today. Having barely begun training its first jet fighter pilots, the young People's Liberation Army Air Force (PLAAF) was forced to meet the challenge of American and Western air power, with the war playing a central role in shaping the new service's strategic and doctrinal thinking, structure, performance and capabilities.

Less than nine months after the end of the Chinese Civil War on the mainland, on 25 June 1950 the leadership in Beijing was startled to hear that war had broken out between North and South Korea.[1] With a U.S.-led coalition, fighting under the banner of the United Nations, intervening days later to repel northern advances into the Western-aligned South, rather than restoring *status quo ante bellum* Washington in September went on to approve a full-scale invasion of North Korea. The invasion was seen to considerably raise the risk of China being forced into the war, with the coalition's Supreme Commander General Douglas MacArthur and many in the Pentagon and State Department strongly advocating using the momentum from the Korean conflict to take offensive action against China.[2] The country had been downsizing its armed forces when the war began,[3] and the PLAAF was at the time far from prepared to provide a meaningful defence.

The PLA had clashed with U.S. forces multiple times during the Chinese Civil War, with Washington having deployed tens of thousands of personnel and considerable air assets for operations on Chinese soil including for offensives against PLA held population centres.[4] The rapid approach of U.S.-led coalition forces to Chinese borders through Korea thus led Beijing to perceive an imminent threat. As China deployed ground units to create a buffer zone protecting key infrastructure on the Sino-Korean border, Chinese personnel and advancing coalition forces first clashed on North Korean soil on 21 October. Hostilities had within a month escalated into large-scale battles involving tens of thousands of personnel from both sides, with all fighting contained to Korean territory.[5]

The PLAAF's first jet fighter unit, the 4th Mixed Air Brigade, had only begun training for its 126 pilots in July, and by the end of September, as U.S. forces advanced quickly across northern Korea, 31 of the brigade's pilots had only just had their first solo flights in the unit's MiG-15 fighters. The unit deployed just 38 of these. Ten weeks of operational training commenced with support from the Soviet 106th Fighter Aviation Division on 1 October, by which time Western forces in Korea had swelled to hundreds of thousands fighting under direct American command.

From the outset, the lack of a viable air force had been a central factor in Beijing's reluctance to deploy ground forces in Korea despite the perceived imminent threat from advancing forces under General MacArthur's command.[6] In the three months preceding Chinese intervention American air power had proven decisive not only for strafing, bombing and drenching North Korean army formations with napalm, but also for targeting logistics to starve frontline units of supplies, which was a leading factor preventing the North from securing a rapid victory. The state of the PLAAF limited Chinese options to prevent its own forces from suffering similar disadvantages, which would significantly impact their morale and operational effectiveness.[7] Other advantages conferred by American air superiority ranged from the ability to expose enemy units at night using parachute flares, which could cause tremendous casualties,[8] to the ability to rapidly evacuate wounded personnel to Japan.[9]

By early 1951 30–40 percent of Chinese supplies were being destroyed in transit by American air interdiction,[10] with approximately 190,000 bombs dropped on North Korea's railways alone during the war – one bomb for every seven metres of rail line.[11] Both combat capabilities and logistics were thus severely restricted

Soviet-built MiG-15 formerly in People's Liberation Army Air Force service. (Wikimedia Commons/Airwolfhound)

MiG-15 at the Victorious Fatherland Liberation War Museum in Pyongyang. (Author's photograph)

in the daytime. Failing supply lines on multiple occasions cut Chinese and North Korean momentum, with armies running out of food and ammunition when on the verge of overwhelming coalition forces and thus forced to suspend their advances.[12] Near the end of the conflict on 10 June 1953, for example, as a last minute Chinese offensive took hill after hill on the frontlines coalition forces were able to respond with massive air strikes – a record 2,143 sorties in one day – to completely halt the advance.[13]

While enemy air superiority imposed a tremendous cost on Chinese forces, it also spurred the rise of China's own fighter fleet which expanded during the war from a negligible fighting strength into the world's third largest. Restricting Western air operations by getting more fighters in the air and strengthening air defences was increasingly seen as key to facilitating more successful ground operations, particularly after massive air attacks on supply lines had played a key role in stalling the Sino-Korean Spring Offensive in April–June 1951.[14] In October 1951, having formed three combat-ready MiG-15 Fighter Air Divisions, the PLAAF began its first significant combat operations in Korea. With new pilots rushed through emergency months-long seven-day-a-weektraining courses, four more divisions became operational in early 1952. Although the rate of the fleet's expansion was tremendous, the PLAAF still remained overwhelmingly outnumbered throughout the war, flying little over 100 sorties daily at its height while Western fleets flew well over 1,000.[15] This disadvantage resulted in a concentration of fighter operations on guarding the northwest corner of North Korea between the Chongchon and the Yalu rivers – an area key to logistical efforts. Although the PLAAF was accompanied by North Korean and Soviet MiG-15 units, the former's scant numbers and the latter's prioritisation of guarding airfields in China itself, and thus their more limited flights over Korea, placed a heavy burden on the Chinese.

The MiG-15 was the PLAAF's first fighter widely deployed for combat operations, and the vast majority of the service's most formative experiences in fighter aviation in its first two decades were gained flying the aircraft and its close derivative the MiG-17. The fighter had become operational in the Soviet Union in June 1949, less than two years after its first flight and just 12 months before the Korean War began. As the first Soviet combat jet considered a world leader in its performance it remains one of the greatest achievements in Russian military aviation history. Having witnessed American nuclear and incendiary bombings levelling 69 Japanese cities in 1945, which in larger strikes could kill over 100,000 residents within hours, the USSR's new fighter was developed primarily to defend Soviet population centres from similar attacks. The perceived imminence of this threat made the MiG-15 a priority program.

U.S. Director of Central Intelligence and Second Air Force Chief of Staff General Hoyt Vandenburg conceded after the first major air-to-air engagements with MiG-15s in Korea: 'Soviet technicians have mastered the design and production problems of extremely high-speed aircraft to a degree which equals and in some respects excels all that we are able to demonstrate in warfare at the present time.'[16] United States Air Force (USAF) Lieutenant Colonel Earl J. McGill, an expert and author on the air war in Korea, concluded: 'the aircraft proved reliable, easily maintained, and more than capable of beating the best the West put up against it.'[17] Regarding its ability to tackle the Western world's top fighter, the American F-86 Sabre, he added that 'the MiG's superior performance left the 86s far behind' and that 'the MiG-15 was a machine well ahead of its day.'[18]

Upon being deployed for operations over Korea, the MiG-15 left all Western fighters entirely obsolete other than the F-86 – an aircraft which formed only a very small portion of American units. The Sabre had a similar swept wing design to the MiG-15 and was rushed into production on a larger scale when it became apparent how overwhelming older fighters' disadvantages were. Even the F-86, however, was far from a full match for its Soviet rival. The MiG-15bis' 0.60:1 thrust/weight ratio was some 30 percent higher, while its more powerful VK-1A engine allowed it to sustain a turn longer and losing less altitude and to quickly climb back to heights at which it was immune to any Western fighter's attacks.

The MiG's service ceiling was around 1,000m higher, and on early variants in particular lighter wing loadings provided a tighter and faster turn rate. MiG pilots frequently waited at high 'perch' altitudes beyond reach before Western fighters reached the operating area, and used their altitude advantage to dictate the time, place and approach angle at which engagements were initiated. MiGs often made diving firing passes before climbing fast to return to high altitude, repeating this consecutively, with Western fighters' much slower climb rates meaning they could not keep up. The MiG-15's superior climb rate on many occasions saved its pilots.[19]

F-86 Sabre. (USAF/J.M. Eddins Jr.)

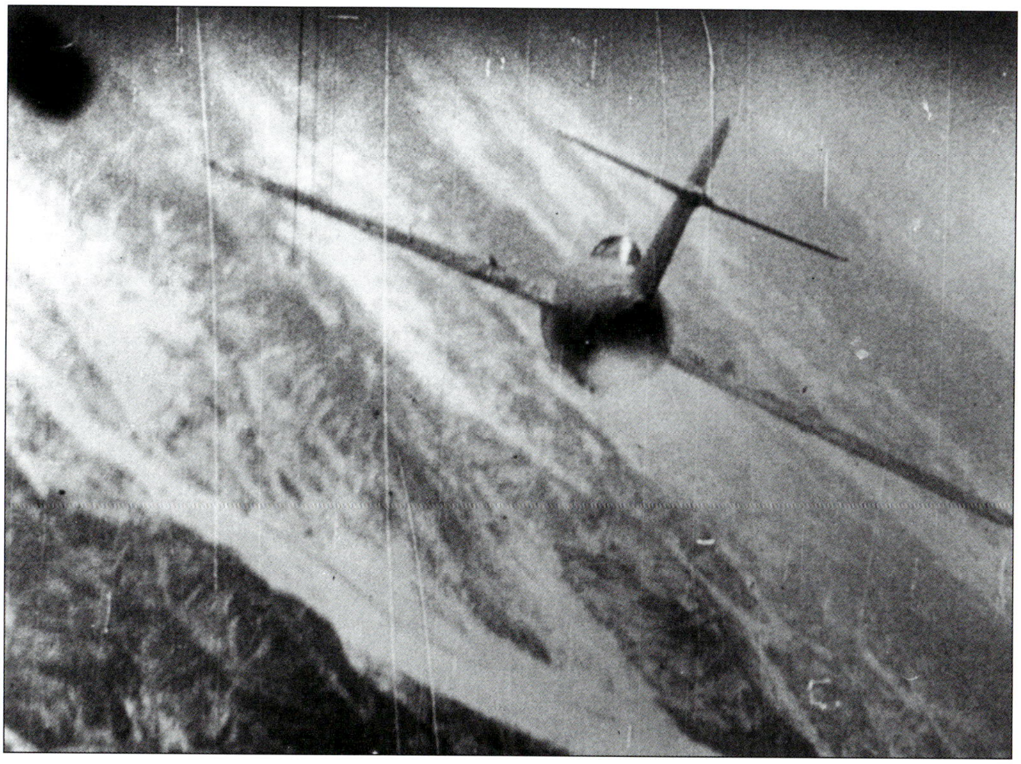

MiG-15 in combat with an F-86 over Korea. (Wikimedia Commons/National Museum of the U.S. Air Force)

Where Chinese personnel were far less well armed on the ground, but were considered hardier and more skilful fighters,[20] in the air much the opposite was true. America's vast pool of veteran pilots, enlarged further by the Canadian and British exchange pilots, flew F-86s with levels of experience which did much to compensate for the MiG-15's superiority. The few pilots the Chinese PLAAF did have often experienced just 80 hours in the air and as little as 20 in jet aircraft when sent into combat, with General Wang Hai, who flew under PLAAF 3rd Fighter Aviation Division in Korea, stating in his memoirs that the average PLA pilot had under 57 flying hours when sent into combat.[21] The tremendous advantages the MiG-15 provided were vital to allowing such airmen to still pose a threat in combat.[22]

Expanded operations from October 1951 resulted in fast mounting successes for the PLAAF, with one of the most notable victories won on 23 October at the Battle of Namsi. There of nine B-29 strategic bombers deployed, MiG-15s shot down three and badly damaged five others – two of which never flew again. This and other victories were sufficient to force B-29s and smaller B-26s to operate only at night from then on,[23] a feat it was noted German and Japanese air power had never come close to achieving.[24] With the B-29 representing a prime symbol of American power, having gained a fierce reputation after its mass incendiary raids against dozens of Japanese cities and nuclear strikes on Hiroshima and Nagasaki, American journalist I. F. Stone was among those to highlight that 'the Battle of Namsi and its aftermath represented a military, technological, and strategic setback of the first magnitude.'[25]

Amid ongoing losses in the air, on 21 November the U.S. Director of Central Intelligence and Second Air Force Chief of Staff General Vandenburg stated: 'a significant and, by some standards, even sinister change has occurred... Almost overnight China has become one of the major air powers of the world... the air supremacy upon which we have relied in the past is now faced with a serious challenge.'[26] From September to December U.S.-led coalition forces lost 423 aircraft over North Korea compared to just 336 by their adversaries.[27] By the end of 1951 some American flyers in Korea considered enemy pilots to be as good as their peers.[28] As one F-86 pilot from the 4th Fighter

F-86s over Korea in 1953. (Wikimedia Commons/U.S. Air Force)

Interceptor Group stated to this effect regarding who was piloting the MiGs and how well they were flown: 'for all we know there may be some ex-American World War II fighter jockeys up there.'[29] This indicated that discrepancies in pilot experience were being heavily compensated for by the MiG-15's advantages.

Chinese and allied successes in the air continued into 1952, and in February America's greatest ace, F-86 pilot George A. Davis Jr., was shot down and killed by a PLAAF MiG-15. This had a serious impact on morale: if the top pilot flying by far the West's most capable fighter could be killed in air-to-air combat then anyone could. Despite having downed 14 enemy aircraft, Davis had written to his wife lamenting very high losses among pilots, stressing that 'things can't go on like they are' and that Soviet-built MiGs 'are so much better than the Sabres.'[30] As the USAF introduced the faster and more capable F-86E and F-86F variants in 1952, PLAAF units were able to retain their advantage by introducing enhanced MiG-15bis fighters,[31] with *Associated Press* reporter Charles Hanley observing: 'more than ever it becomes a duel of technological advance and production capacity, as much as a test of pilot skills.'[32]

Although unrivalled in fighter-on-fighter engagements, the MiG-15 had been primarily designed to neutralise enemy bombers which remained its main mission in Korea. The McGraw-Hill correspondent observed that the MiGs could perform a 'high-speed downhill pass that is virtually impossible to stop,' leaving bombers highly vulnerable. 'MIGs coming downhill can't be headed off even if the Sabre pilots see them at maximum visibility. By the time the Sabres turn 90º to meet the attack the MIGs pass and are on the bombers,' it concluded.[33] Their cannons' greater range and impact power allowed MiG-15s to neutralise B-29s from well outside their defensive ranges.

The successes of Chinese air power had significant strategic implications far beyond the Korean Peninsula, with the *New York Times'* esteemed military editor Hanson Baldwin writing in December 1951 that the MiG-15 had 'forced some of our most ardent advocates of strategic bombing to take another good hard look at our prospects.' He stressed the fact that Soviet-built fighters could devastate American bomber units left the viability of Western bombing raids on major cities in serious question.[34] Journalist I. F. Stone at the time highlighted that the MiG's performance 'showed that the subsonic bombers on which the American military had depended for delivery of the atom bomb in a future war against Russia were indefensible ... America's famous atomic strategy had become obsolete... The subsonic bomber proved obsolete over Korea and with it the foreign policy based upon the subsonic bomber.' Thus, as a result of the MiG-15's performance in Korea it became clear that America could not use its nuclear forces to 'dictate terms.'[35] Assessments in the *Washington Post* were among many to reach the same conclusion.[36]

Western scholars would years later increasingly acknowledge that there was no 'walkover' in the skies of Korea, and that American pilots 'probably highly inflated' their victories.[37] While Western sources initially claimed kill ratios as high as 7:1 (enemies: friendlies), this was later revised to 2:1 – a close fight considering the vast

Memorial statue depicting MiG-15 fuselage at the China Aviation Museum. (Author's photograph)

discrepancy in experience and the nature of the MiG's primary target which were heavy bombers rather than fighters.[38] Chinese sources claimed a victory ratio of 1:1.42 in air-to-air combat over Korea,[39] with other sources pointing to a 1:1.46 ratio in the eight months from September 1951 and a 1:1.42 ratio only after May 1952.[40] In 1952 Chinese pilots reportedly engaged F-86s 85 times, winning 24 and losing 27 engagements,[41] while against fighters other than the F-86 Chinese MiG-15s that year won 30 and lost just two engagements.[42]

AFTER KOREA: CHINA'S PRESSING NEED FOR AIR SUPERIORITY

During the Korean War, China's senior political leadership deeply involved itself with the air force's development and personally took part in related negotiations with the Soviet Union. The PLAAF received sufficient Soviet equipment to outfit 22 air divisions throughout the war and through the first half of 1954 – including ten divisions' worth in 1951 and then one-and-one-half divisions' worth per month until mid-1954.[43] As the newest and by far the most technologically advanced branch of the PLA, with a state of the art inventory contrasting sharply with those of the Ground Forces or Navy, the Air Force gained considerable prestige and was seen to symbolise the new People's Republic's aspirations to strength and modernity. Where the performance of Chinese ground units against overwhelming disadvantages in firepower was described by leading military historians as 'triumph not merely for the prestige of Communism, but for that of an Asian army,'[44] the PLAAF's performance demonstrated that a country for a century derided as backward and unstable could now fight at a world-class level with highly complex weaponry.

The discrepancy between the cutting-edge fighter fleet and the remainder of the military reflected the PLAAF's start from a clean slate in November 1949, its relatively small size, and the great importance attached to it. It also reflected the USSR's willingness to provide its very latest weaponry, as not only were fighters flying over northern Korea far more difficult for an enemy to capture for study than ground equipment, but demonstrating the MiG's prowess did much to ensure Soviet security as well as that of China. This was particularly vital since the Pentagon had been planning for possible nuclear strikes against Soviet cities from as early as 1945, with plans drafted that year involving dropping 204 nuclear bombs on 66 cities.[45] FBI archives revealed that Britain had persistently lobbied Washington to consider such attacks.[46] With the B-29 and similarly vulnerable B-36 responsible for delivering nuclear strikes, demonstrating the former's high vulnerability had significant strategic benefits.

The importance of air defence was also far from lost on the Chinese leadership, not only because enemy control of the skies had heavily shaped the course of the war against Beijing's favour, but also because of the sheer damage done across Korea by Western strategic bombing, for which China itself could be the target in the next war. Bombers targeted 'every means of communication, every installation, factory, city and village' in North Korea,[47] with journalist Robert Jackson reporting that B-29s were fully loaded 'to burn the

American B-29 Superfortresses bombing Korea. (Royston Leonard)

The Chinese leadership's interest in building a strong air arm was prominently addressed in the 1953 First Five-year Plan for the country's military development, which called for expanding the air force to 150 aviation regiments fielding more than 6,000 aircraft and the construction of 153 new airfields. Although this plan was revised to a more realistic schedule after the war, the PLAAF continued to be prioritised for funding throughout the 1950s.[56] China's air doctrine was closely based on that of the USSR with an overwhelming focus on air defence.

Another motivating factor for investment in combat aviation was conflict with remnants of the forces of China's former ruling party the Guomindang, which from 1949 based themselves on Taiwan and surrounding islands. During the Korean War Guomindang forces had played a central role in American plans for a possible joint invasion of the Chinese mainland,[57] and they subsequently continued to be prioritised for deliveries of cutting-edge hardware including AIM-9 air-to-air missiles, F-101 supersonic fighters, and surveillance aircraft such as the U-2. Taiwan served as a key base for U.S. air operations into the mainland, and alongside frequent extensive surveillance operations,[58] CIA reports specified that these included airdropping of paramilitary units for 'promoting domestic anti-government guerrilla operations' and operations to 'collect intelligence and possibly engage in sabotage and psychological warfare.'[59]

Escalated U.S. operations strengthened the perceived importance of air defence capabilities to Chinese security, with the PLAAF armed with new Soviet S-75 air defence systems and MiG-17 and MiG-19 fighters able to shoot down multiple Guomindang and American aircraft.[60] As U.S. intelligence reports concluded: 'what the penetration flights [over China] did accomplish was to motivate the PRC leadership to more quickly build an air force and create an effective air defense system.'[61] This was largely successful, with China coming to field 'an air force that today ranks as one of the toughest to defeat,' as part of a highly potent multi-layered air defence network that left American aircraft highly vulnerable over Chinese airspace.[62] PLAAF and USAF fighters would further clash after 1965, when American violations of southern Chinese airspace during offensive operations against North Vietnam increased significantly. This led Beijing to abandon its prior policy of only monitoring such violations, and in the year following this decision the PLAAF reportedly shot down 12 and damaged four manned American aircraft as well as downing 20 drones.[63]

selected cities from end to end.'[48] Assistant State Secretary Dean Rusk later elaborated that bombers would attack 'everything that moved in North Korea, every brick standing on top of another.'[49] European journalists observed that there were simply 'no more cities in North Korea... my impression was that I am travelling on the moon, because there was only devastation.'[50] U.S. Air Force General Curtis LeMay reported regarding the air campaign: 'we burned down just about every city in North Korea and South Korea both... we killed off over a million civilian Koreans and drove several million more from their homes.'[51] The lesson was not to be forgotten, with Chinese leader Chairman Mao Zedong's own son Mao Anying having himself been killed by an incendiary attack by American aircraft in Korea.

Against this threat, which could be made even more potent with nuclear warheads, fielding of the most formidable anti-aircraft capabilities possible became a priority for the PLA. During the Korean War the U.S. had on multiple occasions strongly considered using nuclear weapons against Chinese forces or population centres in China,[52] with Supreme Commander MacArthur having persistently called for nuclear strikes including over a dozen on Chinese industrial centres.[53] The Joint Chiefs of Staff similarly repeatedly recommended nuclear attacks,[54] and not only was authorisation given to commanders to use nuclear weapons, but nuclear-configured bombers were deployed to East Asia and reconnaissance flights over airfields in northeast China were initiated to obtain targeting data for potential attacks.[55] Primary factors deterring such escalation included the high risk of drawing into the war the Soviet Union, which fielded strategic bombers, nuclear weapons, and a MiG-15 fleet tens of times larger than China's, as well as the demonstrated vulnerability of B-29s over Korean skies, the unexpectedly strong performance of Chinese ground units, and China's expected ability to sustain a fight even if major cities were destroyed.

CHINESE MILITARY AVIATION IN THE COLD WAR

Following the end of the Korean War large acquisitions of Soviet aircraft were vital to facilitating the modernisation and expansion of the PLAAF tactical combat fleet. Seven-hundred-and-fifty regular MiG-15s were delivered to the PLAAF from early 1950, followed by 1,500 MiG-15bis fighters from August 1952 into 1954. As part of broader transfers of technologies and industrial support to China, the Soviet Union supported the production of jet fighters in the country. This first materialised with the assembly of some MiG-15bis fighters under the designation J-2, and then with license production of 1,250 MiG-17s under license from 1955–1960 under the designation J-5. 125 MiG-15UTI twin-seat trainers were also built under license from 1956 under the designation JJ-2. These were followed by the supersonic MiG-19, which was produced in China as the J-6 from 1958 after the USSR supported the establishment of production lines and transferred associated technologies.

Fighter jets were only one of several areas where the Soviet Union comprehensively transferred technologies and full production lines to China, with other notable examples including An-2 transports produced as the Y-5, Mi-4A helicopters produced as the Z-5, T-54 tanks produced as the Type 59, and S-75 air defence systems produced as the HQ-1. This came as part of what leading China scholars referred to as 'the largest planned transfer of technology in world history,' and allowed China's negligible technological and industrial bases to make tremendous advances in the 1950s in both the civilian and military sectors.[64] The technology transfers were so comprehensive that after a breakdown in Sino-Soviet relations in the early 1960 and a total cutting off of Soviet support, China was able to continue production and later conservatively improve on advanced Soviet aircraft and weapons designs domestically. Thus, the J-6, which was the last Soviet fighter class transferred, would remain in production with incremental improvements until 1984, although by this time the lack of access to more up to date Soviet technologies had left China's fighter inventory technologically 25 years behind the Soviets and Americans.

After demonstrating a strong edge over Korean skies, top Soviet fighters would retain their advantages over their Western counterparts into the 1960s, with U.S. analysts summarising regarding the performances of the F-86's successors from the second generation the F-100 and F-104s against their Soviet rivals: 'both were thrashed... over the Taiwan Straits and Vietnam.'[65] Indian Air Force MiG-21s similarly won overwhelming victories over Pakistani-flown F-104s in 1971 – their direct American counterparts.[66] Moving into the third generation of jet fighters from the mid-late 1960s, however, the Soviets began to lag behind in bringing new technologies to the field. Under the administration of Chairman Nikita Khrushchev a strong deemphasis on tactical assets, in the belief that they were largely irrelevant in the era of nuclear and missile warfare, served to seriously undermine the industry's standing.[67] The clearest example was that the MiG-21 which entered service in 1959 but had no viable successor for over a decade until improved variants of the MiG-23 began to enter service in the mid-1970s. During this time clients ranging from North Korea and North Vietnam to Egypt and Syria suffered consequences of being unable to match the top American fighter of the third generation F-4D/E Phantom.[68]

While the Soviet tactical aviation industry was reinvigorated following Chairman Khrushchev's deposition in 1964, it took until the early 1980s and the introduction of fourth generation fighters for the advantage in the air to be regained. The MiG-31 interceptor commissioned in 1981 introduced the world's first, and for two decades the only, phased array radar built for air-to-air combat, as well as unmatched top and cruising speeds, a totally unrivalled ability to sustain Mach 2+ cruising speeds, and an entirely unique capability to simultaneously track and fire on targets both in space and skimming the ground. The MiG-29 medium-weight fighter commissioned from 1982 meanwhile set the trend for widespread use of infrared search and tracking (IRST) systems and introduced unrivalled manoeuvrability, the world's first high-off-boresight air-to-air missile attack capability enabled by helmet-mounted sights, and a formidable austere airfield capability optimising it for deployments near frontlines.[69]

It was the Su-27 Flanker heavyweight fighter however, developed as a higher-end counterpart to the MiG-29 and commissioned from

MiG-15 at the China Aviation Museum. (Military Watch Magazine)

Soviet Su-27 Flanker from the 159th Guards Regiment in Poland. (Rob Schleiffert/Wikimedia Commons)

1984, which would epitomise Soviet leadership in tactical combat aviation in the superpower's final years. A key requirement during development was for the Flanker to outperform the heavyweight F-15 Eagle, which had succeeded the F-4D/E and was by a significant margin the most capable fighter in any Western air force.[70] The Flanker's relative fuselage cross section was 20–25 percent less than on the F-15C and its lighter counterparts the F-16A and F/A-18,[71] while its blended body-fuselage layout generated greater lift with a smaller wing area and thereby reduced the total wetted area, saving structural weight and generating less drag.[72] It had a higher thrust/weight ratio than any Western fighter,[73] while its integrated statically unstable configuration provided outstanding manoeuvrability allowing for altitude changes that were impossible for other fighters.[74] The result was a tremendous discrepancy between what the Su-27 and Western fighers could do at low speeds, as demonstrated at multiple air shows, which raised widespread doubts about the ability of the F-15 and others to compete.[75]

Where the Soviet Union and the United States were competing neck-to-neck in a league of their own to field the top fourth generation fleets, and by 1980 both had fifth generation aircraft well under development, China was still producing MiG-19-derived J-6 fighters that dated back to the mid-1950s and attempting to produce an indigenous derivative of the MiG-21. By the time the Cold War ended, approximately 75 percent of its fighter fleet was still comprised of J-6s, with the remainder split between older J-5s based on the MiG-17 that had first flown before the Korean War, and a small

Su-27 Flanker. (Wikimedia Commons/ Vitaly V. Kuzmin)

Su-27UBK Flanker in Chinese service. (China Military Online/Liu Hang)

number of MiG-21-derived J-7s. Although limited technologies acquired from Israel, France and the United States in the 1980s had provided some improvements in areas such as avionics and missiles, China's domestic research and development (R&D), combat aviation industry and the PLAAF's training and inventories were all decades behind the cutting edge.

China's position in tactical aviation began to recover only after the end of the Cold War, when improving ties with the Soviet Union saw the beginnings of renewed cooperation and the first sale of combat assets in decades – namely a contract for 24 Su-27 fighters signed in December 1990. While the USSR had not offered the Flanker abroad, instead pressing the PLA to accept the widely exported MiG-29, the need to gain export revenues and improve ties with Beijing, and the low possibility of the PLA compromising the technology with transfers to the West, led it to agree to the sale. China was the only country to ever receive Su-27s from the Soviet Union, with three delivered before the superpower disintegrated in December 1991 after which deliveries were completed by Russia within a year.

Upon receiving its first Su-27s China became by far the lowest income country to field heavyweight fighters from the latest generation. While it could not afford a large fleet of such aircraft, a qualitative performance edge over the fleets of potential adversaries was highly prized and favoured over modernising more squadrons with cheaper fourth generation fighters like MiG-29s. The Flanker's unrivalled 4,000km endurance, designed for escort missions deep into British airspace from bases in Poland with a high degree of independence from ground control, combined with a powerful sensor suite and long-range missiles, allowed each unit to patrol large areas of Chinese airspace.[76] The proportion of the Su-27's internal capacity allocated to fuel carriage was greater than any other fighter, 50 percent more than the F-15, and it had an unprecedentedly high mileage per gallon at all altitudes.[77] Specific fuel consumption, the ratio of the mass fuel consumption to the break power, was far lower than any Western rival – 25 percent lower than the F-15 and 40 percent lower than the F-16.[78]

The Flanker was the world's first fighter with load-bearing airframe components made of high-strength titanium automatically arc-welded in a liquid medium; with structural components made of high-strength titanium sheets manufactured by automated dual-arc welding; and with welded titanium structural components advocated using new fixed-temperature annealing technology.[79] Tailslide manoeuvre showed a high level of tolerance to engine intake disturbances, and the fighter could achieve angles of attack far beyond any Western fighter and exceeding 90º.[80] Between 1986 and 1988 the Su-27 had claimed a total of 27 time-to-height and sustained altitude records.[81]

Technologically the Flanker was three decades ahead of anything the PLAAF had previously fielded, with nothing in its fleet being remotely comparable in endurance, situational awareness, manoeuvrability, weaponry or any other metric of combat performance. Integrating the Flanker into the fleet thus provided experience critical to helping the PLA adapt to modern fighter operations, with notable examples including the aircraft's optimisation for beyond-visual-range combat and its ability to operate without heavy reliance on ground control. For the PLAAF and its personnel adapting to the entirely new way the Flanker allowed them to fight in the air, to the multi-generational leap which its new capabilities represented, and to operating a much more maintenance intensive aircraft well over three times the size of their former top fighter the J-7, was a daunting task.

When the Cold War ended the Su-27 was widely considered by both Russian and U.S. sources to be the most capable fighter operated by any air force in the world in terms of its air-to-air performance.[82] In 1995 head of the USAF Tactical Air Command General Joseph Rallston informed Congress regarding the deficiencies of the U.S. fleet in the face of the Flanker: 'we do not need an intelligence effort to realise the Su-27's ascendancy over the F-15 in agility and power.'[83] USAF Chief of Staff General Michael Ryan observed in 2002 that

Chinese People's Liberation Army Air Force Su-27UBK Flanker. (China Military Online/Tang Jun)

if pilot quality were equal Flanker could reliably outperform the F-15.[84] Project personnel working on the F-22 fifth generation fighter similarly stated: 'The Russians are hot fighter jocks and the MiG29 and Su-27 are hotter than anything we have in our inventory,' referring to the latter as 'the top gun at the moment.'[85]

Perceptions of the Flanker's primacy were strengthened by the results of simulated air battles, most notably in 1992 when, at the home of the USAF First Tactical Fighter Wing at Langley Air Force Base, Su-27s proved capable of tackling multiple F-15s simultaneously. They did so even when starting a mock fight at a disadvantage with the Eagles on their tails.[86] With multiple tests revealing similar results, leading expert on Soviet aviation Yefim Gordon observed: 'trying to avoid the adverse publicity for U.S. aircraft, Western media have been tight as a clam about the fact that the Flanker had won in mock combat against the F-15C as witnessed by USAF specialists.'[87] 'As for the Flanker air combat training sessions – both simulated and otherwise – have long since shown that the F-15C loses out in a dogfight even to the baseline Su-27, never mind the Su-30,' he noted, with the Su-30 being an enhanced derivative that had first flown in 1989.[88] Mock engagements with Indian Air Force Flankers in 2004 and 2005 similarly showed an overwhelming disadvantage for the F-15, as did USAF simulated F-15 sorties against Flankers in a complex of 360° simulation games at Boeing facilities in St Lewis.[89] The Su-27's integration into Chinese service was thus a turning point for the evolution of the balance of power in the air in the Western Pacific, which had until then had for decades been growing increasingly unfavourable for Beijing.

AFTER THE COLD WAR: CHINA'S NEED FOR AN INDIGENOUS SUCCESSOR TO THE SU-27

China's investment in Su-27 acquisitions was a result of the PLAAF's strong emphasis on fielding the most capable air superiority fighters possible, in part influenced by the historical memory of the Korean War where, even lacking trained pilots, the new service had had a strong strategic impact due largely to the MiG-15's performance advantages. The Flanker's perceived key importance to China's security was strengthened in 1991 in the final weeks of the Gulf War. American tactical air power during the war was shown to have gained much more widespread applications and play far more decisive roles than it had during the wars in Korea and Vietnam, devastating Iraq's large ground forces, effectively deciding the war's outcome, and in doing so demonstrating that mass mobilisation armies alone were far from able to provide a sufficient defence. As U.S. National Security Policy expert Professor Daryl G. Press noted in a prominent paper for *International Security* in 2001:

> For U.S. foreign policy, the Gulf War seems to show — and the 1999 Kosovo conflict appears to confirm — that air power is now so lethal, and American air power so dominant, that the United States can win nearly cost-free military victories against its foes. For U.S. military procurement debates, the lessons are equally clear: The United States should significantly alter its military procurement priorities to favour air power over ground forces; in the future, decisive battles will be won from the air.[90]

This was a trend widely observed in leading think tanks papers at the time.[91]

America's extreme reliance on air dominance, which was perceived as a precondition for almost all other aspects of its military campaigns, made the ability to blunt or neutralise its air power a key Chinese priority for years to come with even greater perceived importance than in the Korean War.[92] The Gulf War's outcome swayed debate within the PLA further against the Cold War-era 'People's War' concept, and where during the Korean War Chinese forces' emphasis on night fighting had seriously limited American close air support capabilities,[93] Chinese assessments in the 1990s particularly highlighted that new sensors now allowed enemy aircraft to operate with full efficiency at night. Combined with advances in precision weaponry, this was widely seen to have constituted a Revolution in Military Affairs: an irrecoverable change in the conduct of warfare caused by technological advances.[94] The PLA's modernisation drive from 1991, most notably its renewed emphasis on advanced air defence capabilities, largely represented a response to this.

Chinese People's Liberation Army Air Force J-11B. (China Military Online/Yang Pan)

Regarding the impacts of the U.S.-led coalition victory over Iraq in 1991 a prominent Brookings Institute paper concluded: 'The Persian Gulf War marked a new era in American power projection.... Operation Desert Storm showed that U.S. forces could quickly amass at an adversary's doorstep and project power into adversary territory.'[95] As seen in Korea beforehand, the price Iraq paid for losing control of the air was extreme, with an estimated 200,000 Iraqi personnel and tens of thousands of civilians killed in 44 days of hostilities, and well over half a million civilians dying throughout the decade due to the fallout from the bombing of infrastructure, severe malnutrition, and economic sanctions which prevented post-war recovery.[96] This left significant grounds for Beijing to see itself as vulnerable, and with China by 1992 left as one of very few remaining powers outside the Western sphere of influence, prominent figures in both the ruling Communist Party and the PLA increasingly predicted that China or its treaty ally North Korea could be subsequent targets for U.S. military intervention.[97] The perceived threat only grew as the White House came very close to ordering an air assault on North Korea in June 1994,[98] and more so five years later when during a NATO assault on Yugoslavia a B-2 stealth bomber operating on targeting information provided by the CIA dropped a highly precise satellite-guided bomb on the Chinese embassy in Belgrade.[99]

The Gulf War not only strengthened the perceived need for a modern air force, but the need for the PLA and China's defence industries to rapidly raise their technology levels. On 11 January 1991, Central Military Commission (CMC) President and Secretary General Yang Shangkun emphasised that the Gulf War had ushered in an era where wars rather than diplomacy would increasingly settle international disputes. He stressed that new conflicts would be three dimensional – requiring modern air, land and sea elements – and urged the PLA to accelerate research and development efforts accordingly.[100] Cuts to ground forces and increases to overall defence spending were similarly widely called for in the military leadership, with the PLAAF to be the greatest beneficiary.[101]

The CMC held further major meetings in February and March and decided to increase defence spending, with the bulk of the increase allocated to R&D. Finance Minister Wang Bingqian explained increases as follows: 'that the PLA's budget needed more funds to modernize defence in order to handle unexpected changes in the increasingly complicated international arena.'[102] Echoing President Yang's conclusions Defence Minister Qin Jiwei warned the National People's Congress that PLA equipment was 20 years behind that used by U.S. forces against Iraq, and that under the new international environment a larger budget and more R&D spending was vital.[103] Commission for Science, Technology and Industry for National Defence Vice Minister Shen Rongjun similarly stressed the importance of increased spending on high-end technologies, stating that it would be impossible to apply modern tactics without sophisticated equipment.[104]

The *Christian Science Monitor* concluded later in 1991 regarding the shift in PLA priorities that 'the enthusiasm over high technology apparently stems from the decisive role of 'smart' weapons in the Gulf War,' with similar assessments made by other major publications.[105] Thus in sharp departure from the previous two decades military industries and particularly defence-related R&D were given priority over the civilian economy, with military research prioritised with respect to funding in the expectation that it would spin off benefits to the civilian sector, rather than vice versa as before.[106] The new emphasis on R&D would have disproportionately significant implications for the Navy and particularly the Air Force. Manpower was cut by 500,000 over the next five years, which would be followed by multiple further large reductions to ground forces allowing resources to be reallocated largely to the air force.[107]

The Soviet Union's disintegration would significantly bolster the PLA's modernisation drive during the 1990s, with a collapse in domestic orders and in the broader economy leaving post-Soviet Russia's defence sector very heavily dependent on sales to

PLAAF pilots under the Northern Theatre Command march in formation in front of J-11B fighter jets sitting abreast on the flightline prior to a round-the-clock flight training exercise. (China Military Online/Yang Pan)

China, which allowed orders to be met far faster and transfers of substantial highly sensitive technologies to be offered. After becoming the Su-27's first overseas client, the PLA would become the largest client for Flankers by a considerable margin far surpassing orders by the Russian Defence Ministry itself, with two Su-27UBK trainers ordered and delivered in 1992 followed by 24 more Su-27s in 1995 and 28 more in 1998. Under a contract signed in 1996 105 more Su-27s were built in China under license between 1998 and 2007 under the J-11 designation, of which approximately 60 produced from 2001–2007 were enhanced with local technologies as the J-11A variant. An advanced Su-27 twin-seat derivative with an extended range and improved avionics, the Su-30MKK, was ordered in 1999 and again in 2001 with 38 purchased each time, before 24 of the more advanced Su-30MK2 were ordered for the Navy in 2003. China's Su-30 fighters were widely considered the most capable in the world when first received in 2000, and were far more sophisticated than anything the Russian Defence Ministry was willing to finance for its own forces.

In parallel to Flanker acquisitions the PLAAF not only invested heavily in a wide range of supporting assets from new Russian ground-based air defence systems to airborne early warning and control (AEW&C) platforms, but it also revolutionised pilot training. While Russian fighter pilots were often limited to just 20 flight hours per year in the 1990s, Chinese pilots enjoyed a steadily rising number of flying hours which increased by around 50 percent over a decade for Su-27 and Su-30 pilots to approximately 180 hours per year by 2003.[108] These trends continued over the next two decades, and by the mid-2010s American officials were lamenting that Chinese pilots were often getting twice as many flight hours per year as their USAF counterparts – among the very highest levels in the world.[109]

In 1990 the Sino-Soviet agreement on the export of Su-27 fighters to modernise the Chinese fleet was seen to potentially mark the beginning of a new era for the PLAAF, with restored access to top-end Soviet equipment ensuring its top fighter units would again be able to enjoy parity with if not superiority over their American counterparts as they had in the 1950s. Such hopes proved short lived, however, as while Moscow after the USSR's disintegration only proved more willing to supply its most advanced hardware and technologies, it quickly became apparent that the post-Soviet Russian combat aviation sector would no longer be able to compete at a qualitatively peer-level with that of the United States. Russian defence spending by 1998 had contracted by a phenomenal 95.6 percent compared to a decade prior, and the sharp contraction of GDP to little over half of Soviet-era levels, and similarly sharp contraction of civilian industry and R&D, left the country poorly positioned to sustain the Soviet-era momentum that was propelling its combat aviation sector through the 1990s with export financing.[110] Russia's defence sector decline was more acute than spending figures alone would indicate, since R&D and acquisitions, as more flexible parts of the defence budget, suffered disproportionately from the extreme cuts. Thus from 1991–1995 the R&D budget as a proportion of the fast-shrinking defence budget almost halved.[111]

Where the Su-27 and Su-30 had been acquired from 1991–2004 to ensure a performance advantage over the U.S. Air Force's F-15s, as the F-15's successor the F-22 was introduced into the U.S. Air Force from 2005 the Soviet Flanker's continued ability to provide defensive air superiority was brought into serious question. The USSR had begun developing a direct successor to the Su-27 in the late 1970s, and in 1986 the Mikoyan design bureau was approved to pursue full-scale development. A next generation engine prototype for the fighter first flew that year.[112] Designated *Izdeliye* 1-42, and otherwise better known as the MiG 1.42, the program gained investments on an unprecedented scale in areas ranging from its wind tunnel hours to radar cross section (RCS) reduction testing, with Russian sources expressing confidence that the fighter would lead the world in its stealth capabilities.[113]

Soviet MiG 1.44 technology demonstrator for MiG 1.42. fifth generation fighter program. (Pavel Vanka/Openverse)

Soviet planners placed a much greater emphasis on stealth than their Russian successors could, with the country's industrial base and defence budget far better positioned to support a high-performance fighter with world-leading stealth capabilities. Several stealth technologies which did not exist anywhere else in the world were developed for the new fighter,[114] a notable example being a system using whips of plasma from electromagnetic rays generated by the aircraft to absorb radio waves which reportedly reduced the fighter's RCS by approximately two orders of magnitude.[115] The fighter was the world's first with moveable aft portions of its ventral fins, and had an avionics suite well suited to network-centric operations.[116]

The fighter passed the Soviet Air Force's review in 1991 with flying colours, and once details of the design were eventually publicised it was widely predicted to surpass its American rival the F-22, much as the Su-27 had surpassed the F-15 before it.[117] The Soviet aircraft was expected to make its maiden flight before 1994 and enter service in 2001.

The fallout from the USSR's disintegration delayed the first flight of the program's technology demonstrator, designated MiG 1.44, by over six years due to funding shortages, and while tooling had begun in early 1990 to manufacture the MiG 1.42's pre-production airframes,[118] with four prototypes then beginning construction,

Chinese People's Liberation Army Air Force J-11B. (China Military Online/Wan Quan)

none were ever funded to completion. The Defence Ministry concluded that an operationally useful force was unaffordable,[119] and would refocus resources towards the much less ambitious Su-57 fighter, which after sustained delays saw its first squadron formed 23 years after the first MiG 1.42 squadron would have been formed. This marked a turning point not only for Russia, which for the first time since the 1940s would not compete on a peer-level with the United States in fighter aviation, but also for China which lost the opportunity to import a top-end fifth generation fighter. While in 1990 it appeared that the PLAAF could have continued to rely on importing various successors to the Su-27, culminating in a fifth generation fighter in the mid-2000s, by the mid-1990s it was clear that continued reliance on Russian aircraft would have guaranteed that China's top fighter units fell qualitatively behind those of the U.S.[120] America's operationalising of the world's first fifth generation fighter the F-22 in 2005, and Russia's total lack of a rival program by that time, was one of multiple indicators of this.

Where the international standing of Russian fighters diminished significantly from the early 1990s, China's fast-growing economy and even faster-growing defence budget and scale of R&D facilitated its emergence as a leading power in combat aviation by the mid-2000s. This could be clearly observed at the Su-27 production line in Shenyang, where output quality exceeded that seen in Russia due to China's superior machinery for manufacturing and assembly. Advances in satellite communications and navigation, semiconductors, composite materials, missile engines, display screens and a wide range of other areas all did a great deal to press the country's defence sector forward systematically. Chinese industry was thus able to develop a much-improved new Flanker derivative designated the J-11B, which first flew in 2004 with notable improvements including use of the indigenous multimode pulse Doppler radar the Type 1493, a stronger but 700kg-lighter airframe resulting from a modified structure, and much greater use of composite materials.[121] Signs of Chinese leadership in fighter aviation would emerge much more clearly the following decade, and while in the 1950s its then cutting-edge fighter fleet had relied heavily on Russian technologies, in the 2000s China was doing cutting-edge research and development work and pioneering technological advances domestically.

China overtook Russia in defence spending in 1994–1995, and by 1998 had double its defence budget while reaching close to triple by 2009. It exceeded 400 percent of the Russian budget in 2018.[122] This was achieved despite the far more militarised state of the Russian economy, which spent over twice and very often more than three times as great a percentage of GDP on defence. While China's economy was several times as large as Russia's, the discrepancies in industry and high-tech were far greater still with Russia largely evolving into a rentier economy heavily reliant on exports of primary goods such as fossil fuels.[123] Chinese industry was thus competitive in a much wider range of strategic technology areas with key applications in fighter aviation, from semiconductors to machine tooling, where post-Soviet Russian capabilities were negligible by comparison.

Although China independently initiated work on a fifth-generation fighter to succeed the Flanker before the USSR's disintegration, Russia's diminishing ability to provide peer-level fighters to those in the United States gave this program a higher level of importance in the 1990s. The complexity of fifth generation aircraft and their requirements for a very wide range of cutting-edge technologies, many of them with dual uses and requiring a robust civilian tech sector and industrial base, meant China even by the late 1990s had appeared to be in a far better position to develop them than Russia was. Rapid Chinese economic growth and expansion of high-tech R&D in the following decade only widened this discrepancy. The new fighter was intended from the outset to provide a superior counter to the F-22 just as the Su-27 had for the F-15, and as the MiG-15 had years before done for the F-86.

2
THE J-20 FROM CONCEPTUALISATION TO SERIAL PRODUCTION

INDUSTRIAL ADVANTAGES AND FIFTH GENERATION FIGHTER AVIATION

The scale and sophistication of a country's research and development and its industrial base have throughout history been key determinants of the complexity and quality of the armaments it can produce, with this being particularly evident for higher-end more complex asset types such as fighter planes. Moving into the twenty-first century the importance of civilian high-tech industry to defence has only grown as more key defence technologies have been derived from advances in the civilian sector rather than vice versa as was previously more common. As among the most complex and technology-intensive products to develop and manufacture, both the ability to develop and produce fifth generation fighters and the efficiency and cost effectiveness with which this can be done thus closely correlates with the state of a country's industry and tech sector. The phenomenon that is the J-20 program, and its standing relative to its rivals in the United States, therefore cannot be fully comprehended without a brief assessment of the predominant trends shaping both the defence sectors and the broader tech sectors and industrial bases in China and America.

China by the early 2020s accounted for 35 percent of global manufacturing output,[1] boasted the world's largest economy – one-sixth larger than that of the United States[2] – and was outpacing the U.S. in 84 percent of technology research areas with critical security applications.[3] Chinese entities by 2020 were gaining leads in a growing number of tech areas and filing nearly 50 percent of all patent applications worldwide – over twice as many as those in the United States – with this proportion continuing to expand quickly.[4] The country's technological lead was increasingly widely acknowledged from the late 2010s by experts and policy think tanks in the West.[5] Business leaders from multiple Western firms which favoured China for the manufacture of high-tech products despite its high labour costs consistently stressed that its industrial workforce had unique advantages over the rest of the world, particularly in terms of their skill levels and education, with the *New York Times* summarising that the country 'provided engineers at a scale the

Director of the 611 Institute and the J-20's chief designer Yang Wei. (Renmin Zhibao)

F-22 Raptor. (Wikimedia Commons/Rob Shenk)

United States could not match.'[6] China could assemble the skilled personnel required close to 20 times as fast, with their 'flexibility, diligence and industrial skills' having 'so outpaced their American counterparts.'[7] This prevalence of manufacturing and engineering talent was particularly vital due to the very high skill levels needed in complex manufacturing. Production of fifth generation fighters was no exception, and in the United States skilled labour shortages were consistently cited as the leading cause of manufacturing bottlenecks impeding the deliveries of such aircraft.[8]

A notable example of a key emerging field where Chinese industrial and technological leadership directly translated into advantages in its fifth-generation fighter program was digital assembly and testing, which allowed aircraft designs to be flown, modified and upgraded virtually without the need to build real world test airframes – thus providing the potential to significantly shorten development times. The J-20's chief designer, Yang Wei, observed as early as 2018 that in recent years China had adopted this technology to facilitate paperless design and production of multiple new aircraft including the J-20, which shortened R&D cycles and improved product quality. He added at the time that high levels of integration with civilian industry had played an important role, highlighting as an example that 'The carbon fibre and metamaterials used in the J-20 were developed by private companies.'[9] In the U.S. by contrast, while senior Air Force officials in the early 2020s began to express high hopes that digital assembly and testing could seriously speed up the country's own fighter programs, assessments quickly grew increasingly pessimistic as their application even for relatively basic non-stealth trainer aircraft proved highly problematic and failed to accelerate development times as intended.[10]

The rise to primacy of Chinese industry and its tech sector across multiple fields in the post-Cold War-era directly corresponded to its ability to see its first fifth generation fighter program through development relatively smoothly and in a fraction of the time of its competitors. American programs' considerable difficulties and delays, and the more extreme delays and limitations of Russian programs, has similarly reflected the industrial decline, and in Russia's case also a sharp R&D decline, which both faced over the same period. Referred to by Yang Wei as a symbol of 'the development achievements of China's aviation industry in the past 40 years,'[11] the J-20's strong international standing has thus been the product of multiple broader converging trends in the tech sectors, defence sectors and wider economies of both China and of its rivals.

Although the United States brought the world's first fifth generation fighter the F-22 into service in December 2005, by the early 2020s just three other fighter classes of the same generation had entered service worldwide – the American F-35, the Chinese J-20 and the Russian Su-57. Of these only the F-35 and J-20 were both in production and operated in full strength squadrons, with F-22 production having long since ended while the first full Su-57 unit was expected to be completed only in 2024. The immense technological requirements to develop such fighters meant that, despite their distinct performance advantages, very few further programs were expected to produce aircraft capable of operating on a comparable level.

United States government sources defined the following eight features as prerequisites for a combat aircraft to be considered fifth generation:

- use of a stealth airframe with internal carriage of missiles and significant reductions to both radar cross section (RCS) to infrared signature
- active electronically scanned array (AESA) radars
- sensor fusion
- advanced data links and network-centric warfare capabilities
- sophisticated electronic warfare suits with advanced Digital Radio Frequency Memory jammers and electro-optical (EO) defensive systems
- long range multiband EO targeting systems
- advanced glass cockpits

- the ability to reach supersonic speeds without using engine afterburners otherwise known as a supercruise capability

Where many of these features can be seen on newer fourth generation fighters, and were often introduced gradually as Cold War-era designs were modernised, the key feature that distinguishes fifth generation aircraft is the degree to which their RCSs can be reduced. The RCS refers to how much radar energy a body reflects back to its source, with a lower cross section making it more difficult for radars to detect, track or lock onto. While stealth fighters are very far from invisible to radars, particularly those with longer wavelengths, they are detectable at shorter ranges and, most importantly, difficult for radars, and particularly the small high frequency radars on anti-aircraft missiles, to lock on to. The advantages this can provide in all manner of missions, whether for air-to-air combat, air defence suppression, anti-shipping, or even bombing runs with gravity weapons, are tremendous – with even those radar-guided weapons systems specifically designed to target stealth aircraft still doing so with far more difficulty and shorter range limits than they would be able to against a high RCS target.

RCS reductions can be achieved through changes to the airframe shape, as well as through altering the structures of airframe elements particularly contacts between covering, hatches, and moving and fixed parts of the airframe. Use of radar absorbing materials and coatings are a third major means of reducing RCS. Conservative reductions to the RCS of modernised fourth generation fighters have been made in a number of cases, an example being the Russian Su-35S which due to modifications to its airframe has under a third of the frontal RCS of the Su-27S from which it is derived. Multiple fourth generation fighters also make use of stealth coatings and radar absorbent materials, with notable examples including China's own advanced Su-27 derivative the J-16 and the American F-16C/D and F-18E/F fighters. Even with such enhancements, however, these aircraft still have RCSs several orders of magnitude greater than genuine stealth fighters of the fifth generation.

Stealth fighters nevertheless have a number of shortcomings compared to their fourth generation equivalents, with the American F-22 and F-35 in particular having required more maintenance and been much more difficult to keep at high levels of operational readiness. Stealth coatings in particular, especially in the technology's early years, have been time consuming to maintain, and the need to carry weapons internally to preserve a low RCS has also often reduced stealth fighters' firepower compared to their non-stealth counterparts. The F-22, for example, carries up to eight air-to-air missiles while enhanced variants of its fourth-generation predecessor the F-15 carry up to 22 – although this is a more extreme case.

While use of an RCS-reducing stealth airframe has been consistent to all definitions, the other defining features of fifth generation fighters have not always been agreed upon. A notable example is that the American F-35 lacks supercruise which seriously limits its supersonic range. It is also uncertain whether the Russian Su-57 can supercruise. Under a strict reading of requirements this would disqualify them as fully fifth generation fighters. This was also an issue for early variants of the Chinese J-20 produced before 2019. Access to high-off-boresight air-to-air missiles and paired helmet-mounted sights, allowing a pilot to engage enemy aircraft at extreme angles, has also often been considered a requirement, and was standard even in the fourth generation for Soviet fighters from the mid-1980s and for Western ones from the mid-2000s. This and modern sensor fusion and network-centric warfare capabilities are nevertheless absent on the F-22.

A number of factors have meant that there are far fewer successful fifth generation fighter programs than there were in the preceding fourth generation. One is their sheer complexity and the wide range of technologies that all need to be brought to very advanced levels to produce such aircraft. Much as the number of countries that could independently produce successful fourth generation fighters was significantly lower than those which could produce second or first generation ones, most of the countries which developed competitive fourth generation fighters were excluded as serious competitors in the fifth generation. Only the United States and China, tailed a way behind by Russia, are expected to have defence sectors capable of supporting truly fifth generation fighters relatively independently before the first sixth generation fighters begin to be introduced. Even Russia, however, is heavily dependent on overseas inputs, particularly electronics, while the United States also relies on electronics and other key components such as panel displays and circuit boards sourced from East Asia and Europe for the F-22, and more so the F-35. Indeed, the F-35 was pursued as part of a joint program with eight other countries that contributed to R&D and manufacturing. The unrivalled concentration of high-tech manufacturing in China, however, and its far larger scale of research and development, has made its defence sector less reliant on foreign inputs and its fifth-generation fighters the most fully indigenous as a percentage of components used. This position is only expected to improve as the standing of Chinese industry continues to rise relative to leading rivals.

THE J-10 AND THE J-XX: PAVING THE WAY TO THE FIFTH GENERATION

Although China would begin development of a fifth-generation fighter in the late 1990s, in the preceding decade significant work was needed to bridge key gaps between its aviation industry, which was still decades behind the U.S. and USSR, and the fourth-generation fighter production capabilities of the two superpowers. The process of modernising Chinese industry and its tech sector from the 1990s and transforming its military aviation sector was shaped considerably by the development of the country's first indigenous fourth generation fighter – the J-10 – which was also China's first jet fighter that was not derived from a prior Soviet design, making it a major landmark in the industry's development. The program marked a transition from the third world level defence sector of the Cold War-era to one reflecting China's twenty-first century scale of defence spending, high-tech industry and cutting-edge R&D, and was thus key to paving the way for the subsequent development of the J-20 fifth generation fighter.

The J-10 program catalysed development of China's research, development and acquisition process, challenged traditional risk aversion, introduced design competition and linked user needs more closely to existing design and manufacturing capabilities.[12] A top-down decision-making process and vertical management hierarchy was implemented throughout the program, while design improvements were stimulated through introduction of competitive mechanisms between design institutes. China's three main combat aircraft design institutes proposed competing designs for the J-10, signifying a marked change from the past, and between the No. 601 Institute at Shenyang, the 320 Factory at Hongdu, and the No. 611 Institute at Chengdu, it was Chengdu's unconventional double-canard layout that was chosen. The J-10 program catalysed Chengdu's rise to prominence in the fighter aviation industry, where Shenyang had previously held undisputed dominance, which was furthered from the 1990s by Shenyang's focus on license producing

J-10A at Avidarts 2018. (China Military Online/Yang Pan)

and later developing new derivatives of the Soviet Su-27 Flanker. This led Chengdu to be more closely associated with fully indigenous fighters, with experience developing the J-10 being key to allowing it to serve as the prime contractor for the J-20.

The J-10A would join the PLAAF in 2004 with capabilities 'considered roughly on par technologically with America's mainstay F-15s, F-16s, and Navy F-18s,' according to a range of Western assessments.[13] Where the Flankers acquired from Russia over the past 13 years were considered counterparts to American F-15s, the much lighter and less costly single-engine J-10 was broadly equivalent to newer variants of the F-16, allowing the PLAAF to field a complementary high-low combination. Although a major landmark, the J-10's development was assisted considerably by China's extensive access to fourth generation fighter technologies through imports from Russia, with such technologies having already proliferated widely by the 1990s. With the J-20 well under development by the time the J-10 entered service, the new program represented a significantly more ambitious and daunting undertaking as China's access to fifth generation technologies from abroad was much more limited.

In 1997, the U.S. Office of Naval Intelligence identified a Chinese next-generation fighter program which it dubbed in a booklet on potential threat systems as 'XXJ.'[14] Otherwise known in the West as the J-XX, the program was referred to in China as 'Project 718'. Important insights into the program's early goals were provided by a paper titled the *Strategic Study of China's Fighter Aircraft Development*, which appears to have been written sometime between 1996 and 2003 but which became publicly available only in 2016. Written by former vice president and as chief designer at the Shenyang Aircraft Design Institute Gu Songfen, who had served in multiple high-level positions in the aerospace industry including as a leading aerospace scholar in the Chinese Academy of Sciences and Chinese Academy of Engineering, it represents the J-20 program's most definitive publicly available early documentation. The paper was a product of military and aviation industry input and described the rationale for pursuing a heavyweight fifth generation fighter and its requirements and primary roles. It predicted with considerable accuracy how the program would proceed over the next quarter century.

At the time of the paper's writing, U.S. and Soviet fifth generation programs had been under development for close to two decades, although both – and those in Russia in particular – had stalled considerably in the 1990s due to budget cuts and industrial contraction. The U.S. was pursuing two parallel fifth generation programs which became the heavyweight F-22 and the cheaper single-engine F-35, the former designed for air superiority missions and the latter primarily for strike roles. The Chinese aircraft was intended to be able to compete at the same level as the F-22 and comfortably outperform the F-35 in air-to-air combat, with this mission requirement and thus the need for a heavyweight twin-engine airframe making it a direct successor to the Su-27 Flanker. The paper made clear that the program would produce a fighter focused on air superiority and on long range engagements, which was to be capable of neutralising support aircraft such as airborne early warning and control systems, of providing targeting data to friendly assets. The fighter would have an offensive electronic warfare capability and could act as an auxiliary airborne early warning aircraft implying a powerful sensor suite.

The paper emphasised the advantages fifth generation fighters would enjoy over prior generations including stealth, superiority in communications, firepower and information centric warfare. It also stressed the growing centrality of air power in modern warfare. Development of such a fighter, the paper noted, would both further advance China's already fast modernising military aviation industry more broadly and facilitate significant improvements to the PLAAF's capabilities in defence and offense. It highlighted that a 40–50-year life cycle, a RCS of under 0.3 square metres, and a high endurance allowing the fighter to cover all of Japan with one round of aerial refuelling, were key requirements, as was an active electronically scanned array (AESA) radar with a tracking range of 200 kilometres and the ability to track 20 targets. These were all comfortably achieved, although the radar requirements were highly conservative likely in part because they predated the adoption of AESA radars anywhere in the world.

The paper predicted that China's aviation industry would struggle to develop a fifth-generation engine like the F-22's F119 or MiG 1.42's AL-41F, indicating that advanced fourth generation stopgap engines could power initial batches of fighters. It projected full-fledged development would begin around 2006–2007, flight testing around 2013, and service entry around 2020, with the next-generation engine entering service in 2021. Although this schedule was seen as highly ambitious, the fighter would in fact begin flying two years earlier and enter service three-to-four years earlier. Although its engine – the WS-15 – took longer than expected to develop, its power output was significantly greater than levels thought to have been originally projected, giving the twin-engine fighter comfortably more thrust than any other in the world.

By the early 2000s it emerged that there had in fact been two rival fifth generation programs from the No. 611 Institute at

Chengdu and the No. 601 Institute at Shenyang.[15] Chengdu worked on a more radical delta-canard design with two V-shaped tails and lateral DSI intakes, while Shenyang proposed an orthodox 'tri-plane' design with canards, horizontal tails and widely canted tail fins.[16] The 601 Institute's proposed fighter was even heavier and more costly than that which became the J-20, while reportedly accommodating a much smaller radar, with the institute's refusal to guarantee development on schedule further ensuring the success of the rival proposal from Chengdu.[17] As late as December 2002, however, major Western publications such as *Jane's Defence Weekly*[18] and *New Scientist*[19] continued to report that Shenyang would lead development of China's next-generation fighter, which was only decisively disproven years later.

APPEARANCE

In November 2009 PLAAF Deputy Commander General He Weirong said China's next-generation fighter would fly 'soon' and become operational between 2017 and 2019.[20] The fighter made a sudden appearance on 21–22 December 2010, when unofficial images were circulated on the internet showing it conducting runway tests at the Chengdu Aircraft Design Institute's airfield in southwestern China. The fighter's large twin-engine design was a clear indicator that, much like the J-11 and other aircraft from the Flanker series which it was succeeding, it was intended to have a significant range and payload. Its combat radius was later confirmed to be over double that of the F-22 and F-35, allowing it to operate unrefuelled over areas over four times as large, with this discrepancy growing significantly further from the mid-2020s as production aircraft began to integrate the much more efficient WS-15 engine.

At the time of the J-20's appearance the F-22 was still the only fifth generation fighter in service, although multiple issues with the aircraft resulted in a decision in 2009 to cut production to under 25 percent of initially planned numbers. Although initially intended to boast twice the F-15's combat radius, double its sortie rate and half the maintenance requirements, the F-22 turned out to be far inferior to its predecessor in all these key respects.[21] USAF reports highlighted 'maintenance troubles' and 'unexpected shortcomings' with serviceability,[22] with the aircraft failing to deliver improved performance in those areas in which the Air Force assessed the F-15 as most deficient.[23] Although the Air Force had been assured as late as 1999 that the fighter would be easier and cheaper to maintain than the F-15,[24] the fighter's actual maintenance needs were far greater and its annual operational costs at a conservative estimate over twice as high which made a large fleet unaffordable.[25] Although capabilities were seriously watered down and several key systems removed to reduce costs in the 1990s,[26] each airframe still cost 276 percent as much as initially planned at $412 million up from just $149 million.[27] Issues with breakdowns, poisonous effects on pilots' lungs known as 'Raptor Cough' caused by 'a "mosaic" of interrelated cockpit equipment issues,'[28] and problematic avionics and computer architecture,[29] were among several deficiencies which sharpened the contrast with the F-22's famously successful Cold War-era predecessors, the F-15 and F-4.

The state of the F-22 program was a primary factor exacerbating Western concerns when the J-20 made its first appearance, with Western observers widely highlighting that it 'could call into question the U.S. decision to cut funding for its equivalent fighter F-22.'[30] By then it was already far from feasible to reverse the decision on terminating production, however, with plans to destroy the F-22's tooling equipment set to be implemented within

J-20 first technology demonstrator airframe no. 2001. (Chinese Internet)

J-20 first demonstrator airframe no. 2001 conducts first flight. (酒色财气吕洞宾 on Weibo)

J-20 first demonstrator airframe no. 2001. (top81.cn)

J-20 first demonstrator airframe no. 2001 conducts first flight. (酒色财气吕洞宾 on Weibo)

months. The problematic fighters would begin retirement in the early 2020s just a faction of the way through their service lives, by which time availability rates were by far the worst in the fleet, while the closure of production lines for over a decade had made ongoing performance and parts issues more difficult to address.[31] The failure of the F-22 program to provide a viable successor to the F-15, and lack of a long production run with incremental modernisation, removed a key obstacle to the J-20 gaining a dominant standing.

The release of the first images of the J-20 had a considerable impact not only abroad, but also domestically, with expert on Chinese military affairs Richard Fisher being among several to highlight that 'the response within China has been overwhelmingly positive and has spurred national pride to an enormous degree.'[32] While official Chinese sources did not comment on the release, U.S. Director of Naval Intelligence Vice Admiral David Dorsett assured that China was still a long way from fielding an operational fifth generation fighter.[33] Pentagon spokesman Colonel Dave Lapan played down the program as 'not of concern,' stressing that Chinese military aviation was still technologically far behind.[34] U.S. Defense Secretary Robert Gates five days later questioned 'just how stealthy' the fighter really was.[35] Such reassurances were particularly important since, by the year's end, America's sole fifth generation fighter would no longer be in production – with June seeing the last F-22 completed and the mothballing of production equipment announced.[36]

The release of photos of the new Chinese fighter came just days before Defense Secretary Gates visited Beijing to smooth relations, a year after Beijing suspended military ties in protest over American arms sales to Taiwan.[37] Subsequently on 11 January, only hours before Gates met with Chinese President Hu Jintao in Beijing and a week before Hu embarked on a state visit to Washington, the J-20 made its first flight in Chengdu and stayed in the air for around 15 minutes. Official images of a gala reception marking the flight were quickly published with attendees including test pilot Li Gang, director of the 611 Institute and chief designer Yang Wei, and head of the PLAAF General Xu Qiliang. The aircraft would fly again for 80 minutes on 17 April, then on 5 May for 55 minutes when it for the first time fully retracted its landing gear, before demonstrating low altitude manoeuvres on 26 February the following year.

The timing of the flight fuelled widespread speculation regarding its political significance and what message Beijing, or possibly the PLA independently, sought to convey. As the F-22 more than any other weapons system was seen to symbolise U.S. military superiority, so too did the J-20 signify China's own military rise. The *New York Times* thus concluded: 'the symbolism of Tuesday's flight may considerably outweigh its immediate significance.' The *Times* interpreted the timing as miscoordination between the PLA and the civilian leadership which inadvertently sent the wrong message.[38] President Hu and other civilian leaders gave their visitors the impression that they were themselves unaware of the test flight and its timing, with Gates telling reporters that 'the civilian leadership seemed surprised by the test.' He stressed that Hu told him the flight 'had absolutely nothing to do with my visit' which the defense secretary said he believed.[39] 'The Chinese civilians in the room had known nothing about the test... The PLA would pull such a politically portentous stunt without telling Hu,' Gates later concluded in his memoirs.[40] His statement was supported by experts such as Abraham M. Denmark from the Center for a New American Security on the basis that senior officials had no day-to-day involvement in aircraft development and were unlikely to have known about the flight test.[41]

While Western analysts widely predicted China's fifth generation fighter would take another 10 years to enter production, which proved to be over twice as long as China's aviation industry actually needed, Secretary Gates warned that 'they may be somewhat further ahead in the development of that aircraft than our intelligence had earlier predicted.'[42] Chinese military officials at the time had indicated that the new fighter would enter service around 2018.[43] A paper in the *Strategic Comments* journal observed at the time regarding the significance of the fighter's unveiling: 'The project does, however, underline Beijing's aim – also reflected in its land and naval programmes – to develop armed forces that are commensurate with its status as a rising global power... the J-20 should be considered within this broader context, rather than as China's riposte to the U.S.

J-20 technology demonstrator on first flight. (Leaked and Unattributed on Chinese Internet/Guancha)

J-20 second technology demonstrator airframe no. 2002. (机外停车 Rabbit on Weibo)

J-20 first technology demonstrator airframe no. 2001. (Jacksonbobo on Weibo)

Air Force's F-22 Raptor or as an F-35 Joint Strike Fighter "killer."' It concluded: 'It is clear that the West's recent comparative neglect of crewed next-generation combat aircraft has not been mirrored completely by other major powers.'[44]

DEVELOPMENT

Two technology demonstrators were built for the J-20 program designated '2001,' which made the first flight, and '2002' which appeared 16 months later in May 2012. The two were transferred from Chengdu to the China Flight Test Establishment at Xi'anYanliang later that year, where they remained active. The '2002' airframe was subsequently renumbered '2004' and used for initial testing of a unique side bay missile launch system for PL-10 visual-range air-to-air missiles, with some reports indicating it may have integrated a radar. The '2001' airframe was reported to have been used for testing of radar absorbent materials.[45]

The third airframe, which was the first post-demonstrator aircraft and first actual prototype, appeared in October 2012 numbered '2011.' It differed significantly from the two technology demonstrators, with notable changes including redesigned air intakes, a reshaped canopy, an additional inner frame and an Electro-Optical Targeting System (EOTS) resembling those of the F-35, and the addition of stealth coatings. Other notable changes included a retractable aerial refuelling probe, re-profiled vertical stabilisers, smaller underwing actuators, enlarged tail booms housing additional electronic warfare/electronic countermeasures, and self-defence systems such as chaff and flare dispensers to protect the fighter's rear hemisphere.

Prototype '2011' marked the beginning of a new phase in the program leading up to serial production, and over the next two years until the end of 2014 three more prototypes numbered '2012,' '2013' and '2015' made their first flights indicating a very high rate of progress and considerable levels of funding. These were followed by prototype '2016,' which appeared in taxi tests on 11 September 2015 and first flew seven days later. The silver-metallic sheen on the engines of prior models was replaced by a charcoal colour, air intakes had undergone conspicuous changes, and some sources indicated changes had been made to the side weapons bays. The sixth and final post-demonstrator model designated '2017' first flew the following month on 24 November 2015, and building on the changes seen on '2016' it also conspicuously boasted a reshaped cockpit canopy for improved visibility.[46] There were thus eight demonstrator and prototype flight airframes built for the J-20 program before serial production began, compared to 18 for the F-22, 16 for the F-35 and 10 for the Russian Su-57.[47] All J-20 pre-production models made their first flights within a short period of under five years, with full development costs before service entry estimated at just $4.4 billion[48] compared to $26.3 billion for the F-22[49] and $55.5 billion for the F-35.[50]

Alongside the J-20 demonstrators and prototypes, from 2013 a heavily modified Tu-204C airliner served as a dedicated radar testbed for the program at the China Flight Test Establishment. This not only allowed systems to be tested individually, but also tested their integration to automate their working together. Such large aircraft had far more space for onsite monitoring and diagnosis, which would not be possible if sensors were flown on a fighter-sized aircraft. The Tu-204C featured the J-20's front section and radome as well as several electronic warfare and communications systems all housed above the fuselage. Testing of other subsystems was performed by a similarly converted Y-8C transport aircraft. The role of the one-of-a-kind Tu-204C mirrored the use of a modified Boeing 757-200 to support development of the F-22's avionics, with new levels of sensor fusion pioneered by fifth generation fighters increasing the need for such testing including

J-20 first production batch aircraft in yellow primer, image released on National Day, September 2016, preceding the class's service entry. (机外停车 Rabbit on Weibo)

for radars, electronic support measures suites, infrared missile approach warning systems, secure communications systems, data links and other subsystems.[51]

SERVICE ENTRY

Serial production of the J-20 began in 2015, with images of the first serially produced airframe – number 2101 – published in December that year.[52] Beginning production marked first time China's military aviation industry moved into a new fighter generation using predominantly indigenous technologies, and thus represented a major landmark with very significant implications for its future international standing. Like the prototypes and demonstrators, production aircraft were often photographed taxiing and flying in yellow factory primer before being painted. The first flight of a serial production airframe occurred on 18 January 2016, less than two months after the maiden flight of the final prototype. Deliveries to the Air Force began in mid-2016 sometime before August,[53] shortly before the first F-35A first became operational in the U.S. Air Force.[54] Leaving Chengdu at a low initial production rate, the first J-20 unit formally entered PLAAF service in March 2017.[55]

The J-20's entry into service came just six years after its first flight, placing the program entirely in a league of its own for development speed among aircraft of its generation. The F-22 had entered service 15 years after its first demonstrator flight, the F-35 also 15 years, and the Su-57 after 11 years – and 20 years after Russia's first fifth generation technology demonstrator had flown in January 2001. Indeed, even the J-10 had taken six years to enter service after its first flight in March 1998, which when considering how much more complex fifth generation aircraft were indicated that China's combat aviation industry had come a long way in little over a decade. As industry continued to advance rapidly not only were continuous upgrades made to the software of aircraft across the fleet, but successive J-20 batches saw significant improvements affecting both subsystems and stealth capabilities which made each new batch significantly more capable than its immediate predecessor.

The 176th Air Brigade in mid-late 2016 became the first PLAAF unit to receive the J-20 and was based at the Flight Test and Training Base at Dingxin in the Gobi Desert of Gansu Province. The brigade

J-20 with 176th Brigade at the flight test and training base. (九月飞鹰 on Weibo)

J-20 from 172nd Air Brigade with AL-31 engines.(重直风行 Weibo)

almost simultaneously began operating the J-16 and J-10C fighters, which were new iterations of fourth generation designs with avionics and weaponry brought up to comparable levels to the J-20, allowing them to easily network together, employ many of the same tactics and exchange pilots. Assigned to test tactics for new fighters, the 176th was one of two brigades based at Dingxin alongside the 175th which was assigned to tactical training.

Dingxin served as China's only large-scale comprehensive test training base for aircraft and air defence systems, playing a comparable role to Nellis Air Force Base in southern Nevada for the USAF, and frequently accommodated well over 100 aircraft. Benefitting from very wide-open airspace, expansive training range complexes, and clear weather suitable for flight throughout the year, Dingxin's relatively isolated location near the Shuangchengzi missile test range made it ideal for breaking-in new fighters. It also made the base a suitable host location for both the Golden Helmet and Golden Dart competitions which were the most prestigious in the PLAAF.

The 172nd Air Brigade became the second unit to deploy J-20s in February 2018, with the aircraft both delivered and operationalised that month based at the Cangzhou Flight Training Base in Hebei province near the city of Tianjin. The brigade served both as an advanced training unit and a combat reserve unit, and received the fighters just months after it had begun to receive J-16s. A Defence Ministry statement on 9 February highlighted that the deployment marked an 'important step towards fully forming combat capabilities' for the J-20, stressing that its incorporation into its first combat unit would further enhance the PLAAF's comprehensive combat capabilities.[56] Located near China's east coast, the Cangzhou facility was not only one of the closest to the capital Beijing, but was also very close to the Korean Peninsula, with the deployment of J-20s there occurring amid high tensions in the area and shortly after the White House had seriously considered launching an attack on North Korea.[57] After Chinese state media had warned that the PLA would intervene if the U.S. attacked its neighbour,[58] Beijing had drawn a red line with Chinese air defence assets deployed to the Korean border and ships sent to exercise at sea to stress that strikes on its

AL-31 powered J-20 from first serial production batch in late 2016 preceding service entry. (机外停车Rabbit on Weibo)

treaty ally would not be tolerated.[59] The deployment to Cangzhou followed unverified reports that J-20s had flown near or into South Korean airspace as a show of force to deter possible military action, although with the fighter program still in its infancy these were widely dismissed.[60] Beyond Korea, Cangzhou was one of the closest air bases to the capital Beijing, and placed J-20s well within range of targets in Japan and Taiwan. Almost three years after the first J-20s were delivered, in January 2021 the 172nd began to receive more advanced J-20A fighters making it the only Air Brigade to ever operate both variants.

The 9th Air Brigade became the third unit to integrate the J-20 in January 2019, and was based at Wuhu Air Base less than 300km from Shanghai. It was the only frontline unit to receive the baseline J-20 variant before the more advanced J-20A replaced it on production lines in the second quarter of 2019. Otherwise known as the Wang Hai Air Group 'Ace Unit,' it dated back to the 1950s when it flew MiG-15s and MiG-17s and had been responsible for shooting down 59 enemy aircraft – 29 of them American fighters. Its namesake was the Korean War flying ace Wang Hai who was credited with nine kills against American fighters and had led the unit when it was still organised as an air regiment.[61] The deployment was followed by widespread reports linking the J-20 program to the Korean War historical memory of the PLAAF.

Before becoming the first frontline unit to field J-20s, the 9th Air Brigade had also been the first to receive Su-27s delivered from 1991, reflecting both the elite status the unit had long maintained as well as the J-20's status as a direct successor to the Flanker. Both would in their respective eras similarly begin bringing the PLAAF into a new generation. The brigade had been similarly prioritised to become the first to operate the new Su-30MKK from 2000, which was at the time by far the most capable fighter ever produced in Russia and represented the last Russian fighter ever acquired to equip multiple PLAAF squadrons. Although the Su-30 had just a decade prior been considered an elite fighter, by 2019 new indigenous fighters had left its capabilities far behind, leading the Russian jets to be reallocated to lower priority units when J-20s began to be received.

Wuhu Air Base had long been considered an optimal location to stage PLAAF fighter operations in the Taiwan Strait, with the new stealth fighters' close proximity to Taipei being highlighted with such concern by Taiwanese media following their deployment.[62] Research fellow at Singapore's S Rajaratnam School of International Studies Collin Koh was thus among many to speculate: 'The unit turning operational in Eastern Theatre Command is precisely aimed at Taiwan... And to challenge U.S. military activities in Taiwan Strait, besides posing a threat to the median line that Taiwan's air force patrols along.'[63]

The 1st Air Brigade became the fourth unit to integrate J-20 fighters in January 2021 and the first frontline

First image confirming J-20 deployment under the 9th Brigade. (空军发布 on Weibo)

Yang Juncheng, head of the Wang Hai Group. (CCTV Military)

WS-10C engines on a J-20 fighter. (飞扬军事铁背心 on Weibo)

1st Aviation Brigade Commander Senior Colonel Li Ling operates his J-20 fighter jet during the air confrontation training on 10 January 2022. (China Military Online/Yang Pan)

1st Aviation Brigade J-20As taxi in close formation during a flight training exercise on 7 January 2022. (China Military Online/Yang Pan)

unit to receive the new J-20A powered by indigenous WS-10C turbofan engines, which it received near simultaneously alongside the aforementioned 172nd Brigade. Based at Anshan in Liaoning province, it was the first unit under the Northern Theatre Command to field J-20s. The brigade was the first new unit to be equipped with J-20s in two years, as production lines had transitioned from low-rate production with stopgap Russian engines, to a much higher production rate of J-20A variants using indigenous powerplants. With production of the J-20A's new WS-10C engine ending a previous bottleneck, new units would subsequently receive the fighters at a much faster rate.

An assessment at the USAF's China Aerospace Studies Institute saw J-20 deliveries to a second frontline unit as an indication 'that the PLAAF is satisfied with, and confident in, the capabilities of the J-20, and that more combat units are likely to receive the J-20 in the future.'[64] Dubbed the Military Development Vanguard Air Group 'Ace Unit,' which originated from the first aviation combat force of the PLAAF that served in Korea, the 1st Air Brigade represented

1st Aviation Brigade Commander Senior Colonel Li Ling (left) shares experience with fellow airman after completing a flight training exercise on 8 January 2022. (China Military Online/Yang Pan)

an elite unit long prioritised to receive the most capable fighters available, and had previously fielded J-11B and twin-seat J-11BS fighters. All but a few of the J-11s were reallocated when the J-20s were received. The unit was well located to carry out air defence duties along China's east coast, as well as operations into Korea and across the East China Sea.

The deployment to Anshan was considered by a number of analysts to be aimed at influencing the security situation in the Taiwan Strait at a time of high tensions, with Beijing-based naval expert Li Jie being among those to conclude: 'The J-20's new deployment, announced ahead of the Communist Party's 100th anniversary on July 1, is aimed at telling South Korea and Japan that China is strengthening its air defence along the coastal areas, warning them not to join Washington and intervene in the Taiwan issue.' He elaborated: 'None of the J-20s will be deployed near the coasts, because of their 2,000km-plus combat range, which is more than enough to cover the mainland coastal provinces and Taiwan.'[65] Others highlighted that Anshan was in Liaoning province bordering North Korea, and that the fighters could be ideally positioned to provide air cover to China's treaty ally and to signal Washington should it again consider an attack as it had in 2017. In March 2022, when the first J-20 encounter with American F-35s was confirmed to have occurred over the East China Sea, the 1st Air Brigade was considered the most likely to have been involved.

The 5th Air Brigade became the fifth unit to receive J-20s in June 2021, with the unit based southeast of the city of Guilin in Guangxi Province. J-20As replaced older J-10B and J-10AS lightweight fighters,[66] with the J-10Bs having themselves replaced older J-10A fighters in November 2015, and being reallocated elsewhere when J-20As were received. The 5th Air Brigade was thus the first frontline unit where J-20s replaced a fighter class other than Flankers, with the J-10 being less than half its size and having far lower operational costs indicating that the PLA was willing to transition more units from lightweight to heavyweight aircraft to accommodate expanding J-20 numbers. Delivery placed two of the PLAAF's three frontline J-20 units closely adjacent to the Taiwan Strait, with Chinese miliary experts highlighting that bases hosting the fighters were accordingly likely to be priority targets for missile strikes from Taiwan in the event of cross-straits hostilities.[67]

The 56th Air Brigade was confirmed in March 2022 to have become the sixth unit to receive the J-20, with J-20As replacing J-10B and J-10AS fighters at Zhengzhou Airbase, Henan Province, under the Central Theatre Command. Despite the facility's considerable distance from the sea and from major cities, the J-20A's very high endurance and the new variant's supercruise capabilities allowed the fighters to respond to threats as far as Beijing, Shanghai, Korea and the Taiwan Strait. This provided the PLAAF with greater strategic depth, and ensured fifth generation fighters could operate in key areas around the east coast even if airbases in the region were disabled. The deployment highlighted the value of very-long-ranged combat aircraft in an era where major militaries' missile arsenals maintained very wide reaches against one another's airfields, and thus a leading advantage the J-20 would have over its much shorter-ranged Western rivals in the event of a regional war.

In March 2022 unconfirmed reports emerged that a first frontline unit under the Western Theatre Command had received the J-20, making it the seventh PLAAF unit to do so. Other than units at the Dingxin Flight Test and Training Base, the Western Command traditionally had lowest priority for receiving modern fighter classes due to the lack of sizeable U.S. or allied military presences near the region. The largest command by area, it bordered India, Pakistan and Afghanistan and multiple Soviet successor states in Central Asia. Images released on 19 April subsequently confirmed that the 111th Air Brigade based in China's westernmost province of Xinjiang had received J-20A fighters, which replaced J-11A and Su-27 fighters.[68] The 111th had seen multiple changes to its composition over the past 11 years, with very light J-7E fighters replaced by advanced J-11Bs three times their size in 2011, before these were reallocated to other units in 2016 and the brigade instead received the older J-11As and Su-27s which it operated for six years.

Geographically distributing fifth generation fighter units ensured there were opportunities for older units across all theatre commands to train both against such aircraft, as well as in how to operate alongside them. Deliveries of J-20s to new units before older units had received enough of the fighters to fully transition to the class – for example the 5th Air Brigade receiving the fighters before the 1st Air Brigade had completed its transition to the J-20 – indicated that getting more units acquainted with the new aircraft was being prioritised. The Western Theatre Command also housed the bulk of China's expanding land-based strategic nuclear arsenal, which was highly likely to be a primary target for American bombers in the event of a full-scale war and was thus expected to see air defences strengthened as America's B-21 stealth bomber fleet began to grow from around 2030. The B-21 was largely designed with attacks on this arsenal in Xinjiang in mind.[69] Nevertheless, the fact that J-20s under the Western Command were the only ones not within comfortable non-refuelled range to contribute to operations in the most critical potential hotspots, which were all located off the east coast, meant deployments under the command were expected to remain limited.

J-10C under the 131st Air Brigade taxis onto the flightline during a real-combat flight training exercise in mid-September 2020. (China Military Online/ Wang Guoyun)

The 8th Air Brigade became the eighth unit to transition to the J-20 in August 2022. Based at Changxing in Northeast China, it represented the second brigade under the Northern Theatre Command to do so following after the 1st Air Brigade in January the previous year. This followed confirmation of the first J-20 engagement with F-35s over the East China Sea five months prior, likely from the 1st Air Brigade, with the decision to prioritise enhancing units under the strategically located Northern Command potentially having been influenced by such incidents and by the large and growing F-35 presence in the area.

The 55th Air Brigade based at Jining, a central location near the east coast, was reported in March 2023 to have received J-20s to replace its ageing Russian supplied Su-27SK fighters. This remains unconfirmed, although it would be the third brigade under the Northern Theatre Command and the ninth unit in the Air Force overall to integrate the fighters. The central location along the east coast would allow J-20s to comfortably perform air defence duties for most of the country's first tier cities and launch operations across multiple key theatres from Taiwan to Korea.

The 131st Air Brigade based at Nanning under the Southern Theatre Command was reported in June 2023 to have become the tenth unit to receive J-20 fighters, replacing J-10Cs which had been stationed there for seven years after themselves replacing older J-10As in early 2016. Located near the Chinese mainland's southernmost point, the fighters were optimally positioned for air defence duties over southern cities such as Shanghai and Nanjing, as well as the defence of key naval facilities such as Longpo Naval Base – a centre for nuclear submarine operations on Hainan very near to Nanning. The fighters were also very well positioned to contribute to operations in the Taiwan Strait. Despite a significant increase in J-20 deliveries from 2022, information confirming new deployments became increasingly difficult to obtain, and although in 2023 multiple air brigades including the 41st, 55th and 98th, among others, were all reported to have received the aircraft, only two brigades were fully confirmed to have done so by the end of the year - both in the final week of December. The 97th Air Brigade at Dazu Air Base in Chongqing was the first to convert from J-7s to J-20s, which represented one of the most drastic unit conversions not only in China but worldwide both for maintenance teams, who now operated an aircraft almost four times as large, but also technologically and operationally with the two classes being at opposite ends of the spectrum. Located under the Western Theatre Command near Chengdu where the aircraft were built, the new facility was among the most central in China. The second unit was the 4th Brigade based at Foshan Airbase under the Southern Theatre Command, which in contrast to the remote facility at Dazu was among the very closest bases to key economic centres at Shenzhen and Hong Kong, while also being just within 650km of both Hainan and the Taiwan Strait.[70]

AIRFRAME

The J-20's airframe has a smooth surface with no protruding inlets or pitot tubes, which combined with flush hexagonal fuel-air heat exchangers serves to significantly reduce its RCS.[68] Further contributing to stealth capabilities, the airframe benefits from a frameless canopy, chiselled nose section, blended fuselage, chimed forebody and electroconductive canopy, flat fuselage bottom, sawtooth edges on compartment doors, as well as radar absorbent coatings. The embedding of a retractable in-flight refuelling probe under the fuselage enables in-flight refuelling without significant compromises to stealth.[71]

The airframe makes very high use of composite materials at 29 percent of its overall weight, with titanium and titanium alloys making up a further 20 percent and the remaining 51 percent made from aluminium and steel alloys. Composites are vital to reducing the number of fasteners in an aircraft's structure to provide

Closeup image of J-20A with J-16s in the background. (空天砺剑 on Weibo)

J-20 in yellow primer with drop tanks. (机外停车 on Weibo)

J-20 during low altitude flight over northwest Gansu province. (China Military Online/ Yu Hongchun)

thermophysical and radio-technical characteristics meeting fifth generation requirements. Even in the 1990s when they first began to be widely used on fighters, composites could increase an airframe's weight efficiency by around 25 percent. Chinese industry made considerable progress developing composite materials for use on combat aircraft from the 1990s, and development of the J-20's high composite airframe largely built on extensive work done to enhance fourth generation fighters using these materials. Concurrent advances in civil aviation, including manufacturing of a fast-growing range of airliner components, also played an important role in cultivating expertise in composite material fabrication.

Mature RCS testing technologies played an important role in the development of the J-20's stealth profile, and to maintain a reduced RCS the fighter carries armaments internally.[72] A central weapons bay on its underside accommodates four long range PL-15 air-to-air missiles, with two smaller lateral bays behind the side air intakes each accommodating one smaller PL-10 short-range missile. Weapons integration tests at Chengdu for an entirely unique retractable side missile launch rail for the PL-10 began no later than the first quarter of 2013. The fighter's pioneering of a new stealthier launch system relative to its American rivals was cited by analysts as one of multiple important early indicators that the J-20 could improve on existing designs overseas rather than simply bridging performance gaps.[73]

Four external hardpoints on the J-20 can accommodate either fuel tanks or additional missiles, although doing so would compromise the fighter's stealth profile. External pylons can also be jettisoned to enter stealth mode mid-flight. It remains uncertain which missile classes the fighter can carry externally, and whether it can carry the very-long-ranged PL-XX which is too large for its internal bays and was developed primarily to target enemy support aircraft and bombers.

Leading edge extensions, body lift and canards between them increase the J-20's lift by 80 percent, which in turn allows for a reduced supersonic drag through use of a smaller wing while ensuring a high turn performance by avoiding compromises to transonic lift-to-drag characteristics. Chief designer Yang Wei explained the benefits of this design as follows: 'We have created the J-20's "lift body, strake wing, and canard" layout that is unique worldwide, giving the aircraft not only excellent stealth capabilities but also a strong supersonic flight performance and high manoeuvrability, allowing it to fly further and carry a larger payload of weapons.' 'Many breakthroughs and unique techniques' were similarly made to optimise the J-20's situational awareness, information warfare, internal weapons carriage, and cooperative combat capability.[74]

The J-20's canard delta wing design with twin outward canted all-moving fins is entirely unique among operational aircraft of its generation, providing a far greater lift coefficient than its closest foreign rivals and improving flight performance including both low-speed agility and supersonic manoeuvrability. This is particularly useful for air superiority fighters to provide better control over their angles of attack. They also better optimise the fighter to operate from shorter airfields. Canards on the J-20 appear to be optimised for stealth particularly in the forward sector. Chengdu research papers from as early as 2001 discussed the benefits of a delta-canard design, which had been under consideration since the 1960s for the cancelled but highly ambitious J-9 single-engine fighter and were applied to its successor the J-10.

Although claims that the canard fuselage join is not ideal for stealth have been common since the J-20 was unveiled,[75] they contradict studies performed in both China and the United States. A Chengdu Aircraft Design Institute paper, for one, specifically addressed the impact of a canards on RCS in a publication that emerged in 2019 concluding that there was no significant impact on stealth capabilities.[76] Lockheed Martin notably considered a canard delta wing stealth fighter for its Joint Advanced Strike Technology program before settling on the design that became the F-35, while Northrop Grumman also considered a canard aircraft derived from its YF-23 technology demonstrator, indicating that the physics of canards were compatible with stealth.[77] The X-36 fighter agility research aircraft developed by McDonnell Douglas and NASA also used a canard design and was considered among the stealthiest manned jet aircraft to have ever flown.

J-20 makes a left turn. (野生灰熊 on Weibo)

A rare clear top view of the J-20's cockpit. (夜航大叔 on Weibo)

J-20A pilot in cockpit. (空天砺剑 on Weibo)

AVIONICS

J-20 pilots interact with a modern glass cockpit with a 610 x 230mm touchscreen panoramic display controlled by voice command, a smaller liquid crystal display between the pilot's legs, and a prominent wide-angle holographic heads-up display with a distinctive green glow. The fighter is flown with a side-stick and throttle, and a helmet mounted display was confirmed in May 2018 to be operational.

Situational awareness is provided by the Type 1475 (KLJ-5) AESA fire-control radar, the EORD-31 IRST, and an electro-optical distributed aperture system. The cockpit design and helmet mounted display allow the pilot to better process data provided by onboard and offboard sensors and benefit from 'information fusion.' The Type 1475 is very possibly the most powerful radar built for air-to-air combat in service worldwide, and has between 2,000 and 2,200 transmitting/receiving modules compared to 1,900 on the F-22's APG-77, 1,600 on the F-35's APG-81, and just 836 on the French Rafale's RBE-2AA. Its 24kW transmission power compares favourably to the APG-77's between 16.4 and 20kW.[78]

Electronically scanned array radars are not only far more reliable and less susceptible to jamming, but also facilitate a much wider range of electronic warfare techniques, are significantly more powerful and efficient and can scan much faster than those with mechanically scanned arrays. While mechanically scanned array radars rely on a single continuous radio beam, electronically scanned array radars emit multiple intermittent beams from hundreds of discrete individual electronically steered emitters which can narrowly aim at multiple objects to simultaneously track targets, search for other possible targets, track friendly missiles, and send commands to multiple missiles to intercept multiple targets. AESA radars have the further advantage over passive electronically scanned array (PESA) radars of being able to scan more precisely and send out radio waves at different frequencies in multiple directions simultaneously without moving any antennas, which is particularly valuable for electronic warfare. They also have significantly lower signatures making it much more difficult for adversaries to detect their emissions and use these to home in on radar's location. This makes such radars vital to an aircraft's ability to maintain stealth.

Development of electronically scanned array radars for fighter/interceptor aircraft was pioneered by the Soviet Union, with the

A clear image of the hexagonal window from the J-20's distributed aperture system, with pilot Chen Liu in the foreground. (Ministry of National Defence of the People's Republic of China)

Zaslon PESA radar entering service in 1981 equipping the MiG-31 interceptor. It was followed by a much-improved new generation PESA radars introduced the following decade including the Zaslon-M, which was ready for serial production in 1994, and the N011M Bars which completed development at around the same time. The Soviets were followed 19 years later by the United States, which integrated AESA radars onto a modernised F-15 squadron in 2000, followed by France in 2001 with a PESA radar on the Rafale and Japan in 2002 when its F-2 became the world's first fighter with an AESA radar as its standard sensor.

The J-20's AESA radar is complemented by the EOTS-86 optical sensor, an equivalent to the F-35's EOTS, which is well optimised to engagements against stealth targets and provides long range visual identification of potential threats. The EOTS-86 is highly complementary to the fighter's IRST, with the latter detecting and the former identifying targets before they are neutralised at beyond visual range. This can be done passively without the need to emit any radar emissions which allows the fighter to better conserve its stealth capabilities.

Six diamond-shaped windows, with two on the sides of the J-20's nose, two more under the rear fuselage and a further two at the forward aft of the cockpit, house electro-optical sensors covering all directions which form part of a distributed aperture system. It allows pilots to observe their surroundings 360° around the aircraft as well as through the bottom using their helmets, which among other functions helps to eliminate blind spots which is particularly valuable in visual range combat. The system facilitates the detection and tracking of enemy missiles including launch point detection and the cueing of countermeasures, as well as detection and tracking of enemy aircraft and cueing for air-to-air munitions and infrared tracking systems. It also provides imagery for cockpit displays and pilot night vision. The designation of the J-20's aperture system remains unknown, with its only known equivalent being the F-35's AN/AAQ-37. Use of aperture systems is one of multiple examples of a key cutting-edge technology which sets the J-20 and the F-35 apart from other fighters even within their generation.

The benefits of the J-20 and F-35's aperture systems were effectively summarised by Russian defence analyst Vladimir Tuchkov as follows:

> It is the pilot's helmet, which makes the aircraft 'transparent.' That is, visibility is not limited by the cockpit windows. The whole panorama of the surrounding area is displayed in the pilot's visors, in both the visible and the infrared spectrum. Monitoring the pilot's head and eye movements, the computer provides the necessary panoramic viewpoint and provides the pilots with tips, and manages targeting.

He indicated that a lack of such systems was a leading disadvantage for the Su-57 – one which also hindered the older F-22.[79]

ENGINES

As predicted by Chinese aviation industry experts almost 25 years before the J-20 entered service, engine technologies proved to be the most challenging area of development due among other factors to the very high requirements for power output, fuel efficiency, reasonable maintenance, and near perfect reliability.[80] Although work on a suitable engine began in the 1990s, which was later designated WS-15, in contrast to almost all other aspects of the J-20 program which completed development on or ahead of schedule, the engine took considerably longer than expected to be ready for serial production. As such stopgap engines based on powerplants designed for fourth generation fighters were relied on to power the J-20.

China's fighter aviation industry for its first six decades relied either on reverse engineered Soviet turbojet designs from the 1950s, such as the J-7's WP-13 engine based on the Soviet R-13, or on modern turbofan engines imported from Russia from the 1990s. While turbojets were phased out of production in the U.S. and USSR in the 1970s in favour of turbofan engines – which were far more fuel efficient and seen to have much greater potential for modernisation – even among turbojets engines Chinese designs like the WP-13 were considered outdated, particularly by the 1990s. As their name implied, a turbofan engine achieved a superior performance by using a ducted fan to increase the mass and reduce the speed of the propelling jet. Development of an advanced turbofan engine became more viable as Chinese industry modernised rapidly in the 1990s, and as it gained access to Soviet technologies and expertise that decade, paving the way for the indigenous WS-10B's entry into service in 2009. As the first indigenous fighter engine and the first Chinese produced turbofan, its comparable thrust/weight ratio and overall performance to those powering Russian and American fourth generation fighters made it a major milestone for the industry.

Although highly capable as a powerplant for fourth generation aircraft, the WS-10B would have left the J-20 far underpowered, with the PLAAF consequentially commissioning a much more powerful engine for production variants of the J-20 from 2015. The engine was widely thought to be a modified derivative of the Russian AL-31FM2: an enhanced AL-31 variant with approximately 16 percent more thrust designed to power the world's heaviest fighter class; the Su-34. The engine used an entirely new three-stage fan, new high-pressure turbine assembly and combustion chamber, a 75°C higher turbine entry temperature, and a new full authority digital engine control with hydromechanical backup – a major leap in sophistication over prior AL-31 models.

A variant of the AL-31FM2 lightly modified for the J-20 is thought to have been jointly developed with Chinese inputs and had an estimated afterburner thrust of 142kN – just seven percent lower than the F119 powering the F-22. The variant developed for China may well have integrated technologies from the AL-31FM3 engine, which was first tested in 2007 and was considered production

AL-31 engines powering a J-20 fighter at Zhuhai Airshow in 2016. (Jacksonbobo on Weibo)

WS-10C powered by WS-10B engine. (央广军事 on Weibo)

Prototype no. 2021, built to test WS-10C engine integration. (机外停车 on Weibo)

ready around the time Chinese orders were made. It had a significantly superior performance to its predecessor with major improvements to its turbine fan and combustion chamber in particular. Offering these technologies for the J-20's customised powerplant would have provided a means to recoup some of the investment sunk into developing the engine, which never received orders from the Russian Defence Ministry itself.[81] According to figures from the Stockholm International Peace Research Institute, 80 AL-31FM2-derived turbofans were purchased in 2014, which represented one of the last ever Chinese orders for Russian fighter engines.[82]

As the first J-20s became operational in early March 2017 vice president of the Aero Engine Corporation of China Beijing Institute of Aeronautical Materials Chen Xiangbao predicted: 'It will not take a long time for our fifth-generation combat plane to have China-made engines... The engine's development is proceeding well. We also have begun to design a next-generation aviation engine with a thrust-to-weight ratio that is much higher than that of current types.' He elaborated regarding remaining issues in the industry: 'For instance, we are able to develop the two most important components in an advanced engine – the single crystal superalloy turbine blades and powder metallurgy superalloy turbine disks – but in mass production, the products' quality is not very satisfactory.' 'The road to success is filled with setbacks and failures. Each of the world's engine powers has walked this road,' he added. His statements were presumed to be referring to the clean sheet WS-15 design which was at the time expected to replace the jointly developed AL-31FM2 derivative (from here on referred to simply as the AL-31FM2) within five years.[83]

WS-10C engines on the J-20 fighter. (飞扬军事铁背心 on Weibo)

With the WS-15 facing delays, an improved variant of the WS-10 of unconfirmed designation – speculated to be designated (and from here on referred to as) the WS-10C – was developed to provide a viable replacement to the AL-31FM2. The engine was first confirmed to be in flight testing in 2017, with a new fighter prototype designated '2021' built to test its integration and likely other new features for the J-20A variant. Aircraft '2021' was first seen in September 2017. The WS-10C powered all factory-built J-20s from mid-2019. The first J-20s with the new powerplants, designated J-20A and benefitting from a number of conservative other improvements, entered service from January 2021.

The WS-10C's combustion chamber was reportedly an area where it improved on the AL-31FM2 particularly significantly, with the engine also having improved fuel efficiency, acceleration performance and thrust over the Russian turbofan. It also had much better stealth capabilities most conspicuously represented by its serrated afterburner nozzles – a key distinguishing feature from other engines in PLA service. Engines with such stealth features were first seen tested as early as 2014, when they were integrated onto the first prototype number '2011' which was re-engined to serve as a testbed for such powerplants 18 months after its first flight. The WS-10C reportedly allowed the J-20 to fly at low supersonic speeds without afterburners, which allowed it to sustain supersonic flight for much longer periods.[84] This allowed the J-20 to become the only supercruise-capable fighter of its generation in production worldwide.

WS-10C-equipped J-20 fighters had by early 2022 operated at extreme altitudes, humidity levels and temperatures, which was cited as evidence that the fighter could carry out missions across the country. Such testing was thought to be the purpose behind the J-20's reported brief deployments to southwestern China's mountainous territories, where take-offs at extreme altitudes were particularly challenging and required high engine power due to the thin air. This had previously posed similar problems for the Indian Air Force's European-sourced aircraft near its northern borders. The J-20's deputy chief designer Gong Feng highlighted in January 2022 that integration of the WS-10C significantly increased the fighter's capabilities and was customisable for the aircraft, where the AL-31FM2 had limited its performance. Gong had been tasked with conducting reliability tests for the fighter under extreme conditions, and concluded: 'I can proudly tell everyone, that the J-20 with Chinese engines has it all, both from the inside and outside.'[85]

Visiting fellow at the Griffith Asia Institute in Australia Peter Layton highlighted in April 2022 regarding the significance of the WS-10C's integration, which he claimed was more reliable than any Russian engine: 'It's not just that China does not need Russian help anymore, it is that Chinese-built aircraft are now superior to Russia.' The engine's superior performance and reliability relative to the AL-31FM2 made long range maritime patrols 'a much more plausible option' for the J-20, he added.[86]

The J-20's final intended powerplant – the WS-15 – was designed by the Shenyang Aeroengine Research Institute and was to be produced by the Xi'an Aero-Engine Group. It would not only provide an estimated 21–36 percent more thrust than the AL-31FM2 and WS-10C (estimates varied widely) but would also have a much higher thrust/weight ratio estimated at 10:1 or 11:1, require less maintenance, improve thermal management properties, and have a significantly longer service life. Its cutting-edge single-crystal turbine blades, although proving particularly difficult to mass produce, were expected to almost double the engine's lifespan compared to the WS-10B while significantly lowering maintenance needs. The new powerplant was considered broadly equivalent to the F119 powering America's F-22, but had more ambitious specifications with an expected maximum thrust of 18.4 tons,[87] compared to 17.5 tons for the F119, 16.3 tons for the WS-10C and AL-31FM2, and 18.7 tons for Russia's AL-51F then under development for the Su-57.[88] Its weight/thrust ratio was expected to be between 0.09 and 0.1, compared to 0.125 for the AL-31F and 0.11 for the F119, making it more efficient than any prior powerplant used by a twin-engine fighter.[89]

Pre-development and demonstrator work on an engine to power a fifth-generation fighter began in the 1990s, while the WS-15 began as a program in 2006. In 2009 prototypes reportedly achieved a thrust/weight ratio matching the F119. Although its development was protracted by the standards of Chinese weapons programs, the time required was far from exceptional when compared to American and Russian next-generation engines. In the United States Pratt & Whitney and General Electric spent over 12 years just to develop prototypes of the F119, and it took another 14 years of testing to perfect the design for operational service. China's relatively poor technological base when the WS-15's pre-development work started in the 1990s inevitably slowed this significantly further, and with even the country's fourth generation level WS-10B only maturing around 2010, the fact that the WS-15 could begin flight testing in 2016 and enter serial production in 2023 represented a tremendous rate of development. The much more mature stage which China's aeroengine industry had reached by the mid-2000s when the program began resulted in a much smoother development timeline compared to the WS-10, which had begun development in the 1980s when the industry had been very far behind.

Reports emerged in October 2016 that the WS-15 would begin flight testing on the IL-76LL – a heavily modified Soviet transport jet – and had finished ground testing, with multiple successful tests reported over the next five years,[90] culminating in the first images of a J-20 flying with the new engine emerging in January 2022.[91] The J-20 used one WS-15 prototype and one older WS-10 engine, with the latter providing redundancy and enough power to land in case there was an issue with the WS-15. Further reports of an improved test performance emerged two months later.[92] Extensive coverage that year by Chinese state media, and subsequent confirmation that the J-20 was indeed test flying the engine, were seen by a number of

Prototype 2052 first flight with two WS-15 Engines. (飞扬军事铁背心 from Weibo)

analysts as important indications that the WS-15 would enter service in the near future.[93]

In March 2022 head of the Aviation Industry Corporation of China's Manufacturing Technology Institute Li Zhiqiang expressed confidence that complex problems of warplane engine development could be resolved. A specialist in the hot forming and superplastic forming of titanium alloys, Li observed that the technology gap with world leaders had been narrowed and that new developments would provide fighters with higher thrust/weight ratios, thrust-vector control and variable cycles.[94] Days later experts credited breakthroughs in key material technologies such as those in single-crystal blades, the accumulation of technologies and theories over two decades of accelerated development, and the establishment of the Aero Engine Corporation of China, with boosting the jet engine industry.[95] In December a professor at Beihang's University School of Energy and Power Engineering, Liu Daxiang, a leading advisor to miliary aviation programs, strongly indicated in a prominent keynote speech that the WS-15's development was nearing completion, and confirmed a weight/thrust ratio of at least 0.1.[96] Three months later an official from the Aero Engine Corporation of China confirmed that the WS-15 had entered serial production, indicating a low initial rate of production which could allow it to enter service in late 2024.[97] Providing a strong indication of progress in reliability levels, the WS-15 was first flown in twin configuration on J-20B prototype '2052' in June 2023.[98]

The WS-15 was expected to significantly improve the J-20's endurance, flight performance and availability rates, lower its maintenance requirements, and reduce its heat signature particularly from the rear, overhead and underneath. Its greater power would also be key to facilitating use of more powerful new generations of sensors and potentially a range of directed-energy weapons. A significantly higher thrust/weight ratio for the fighter ensured a higher climb rate, greater manoeuvrability at all speeds, faster acceleration and likely a higher altitude ceiling. Thrust-vectoring nozzles would also also very significantly further improve the fighter's manoeuvrability at low speeds. The state-run *Global Times* thus summarised in March 2022 regarding the WS-15's future integration: 'the J-20 will finally become its ultimate form.'[99]

ARMAMENTS

Much as had been the case for the F-22, development of new classes of air-to-air missiles was necessary to fit the J-20's internal weapons bays. In America new variants of the existing AIM-120 and its short-ranged counterpart the AIM-9 were developed with clipped fins and improved overall performances for integration onto the new fighter. Although these missiles had entered service from 1991 and 1956 respectively the need to reduce costs particularly after the Cold War's end led to an emphasis on upgrading older designs and meant

A J-20 with PL-10s in firing position and open main weapons bay showing PL-15s at Zhuhai 2018. (Wikimedia Commons/emperornie)

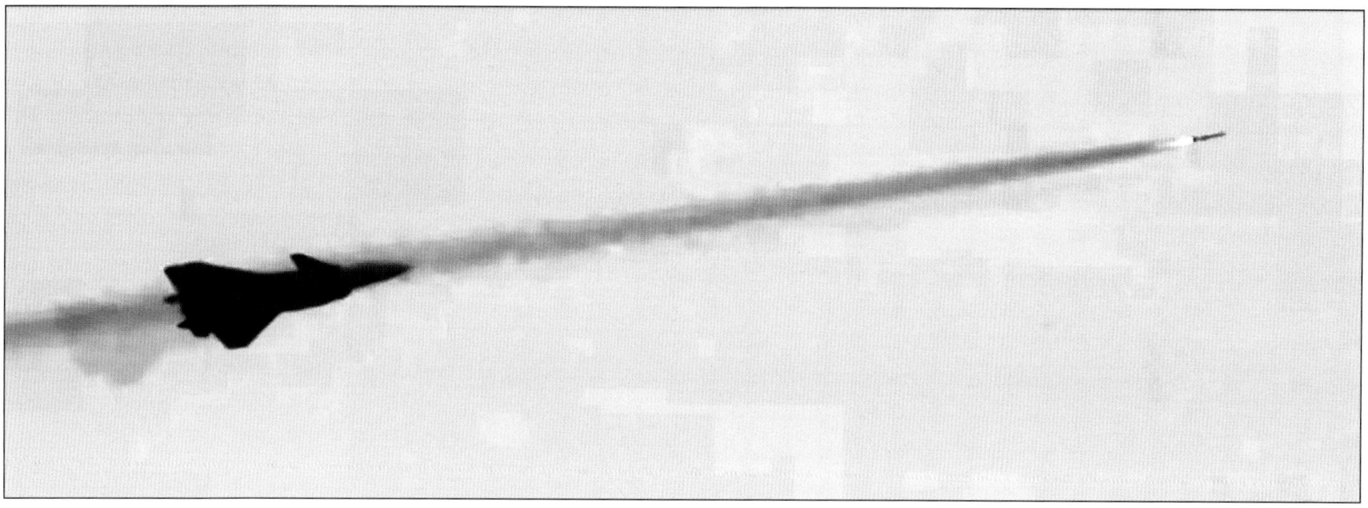

First public image of a J-20 firing a PL-10 air-to-air missile. (太湖军I 名 on Weibo)

neither had an operational successor even by the early 2020s. This contrasted to the PLAAF, which despite having far newer air-to-air missiles in production still developed entirely new ones, the PL-15 and PL-10, for the J-20.

The PL-15, a challenger to the American AIM-120D and European Meteor, had a longer range than its competitors with estimates ranging between 200km and 300km.[100] Designed from the outset for integration onto stealth fighters, with shortened fins to facilitate internal carriage, its use of a dual-pulse rocket motor was key to its range advantage, while a two-way datalink provided guidance updates to both the missile and the launching fighter. The missile notably used an AESA radar for inertial guidance which, as a prominent 2018 International Institute for Strategic Studies (IISS) report noted, ensured 'improved performance against low-observable targets and greater resistance to countermeasures on target aircraft, such as radio-frequency jammers.'[101] Despite some initial speculation that the American AIM-120D introduced in 2014, with an estimated range of 160–180km,[102] might also use an AESA radar, this never materialised.[103] The PL-15 caused considerable concern among Western analysts[104] and U.S. military officials,[105] and its high performance was widely considered a leading reason for the launch of the AIM-260 missile program 2017.[106] Its capabilities, as noted by analysts at *The Drive* as late as 2022, had left the U.S. 'working to close the gap' by developing clean sheet successors to the AIM-120.[107]

The short-ranged PL-10, which used thrust-vector controls and a gimbaled imaging infrared seeker, was capable of high-off-boresight engagements and could engage aircraft beyond 90°, thus providing a significant performance advantage over the vast majority of foreign competitors. Its cutting-edge seeker type provided a range of significant benefits including superior resistance to countermeasures and an improved detection range. According to a report by IISS its induction 'has placed China among the handful of nations with a defence-industrial base capable of producing such a weapon,'[108] with analysts at *The Drive* highlighting that it was 'judged to be at least equivalent to Western weapons in its class.'[109] Reports from the British

A closeup of the side weapons bays as the J-20A fighters taxi. (空天砺剑 on Weibo)

J-20As with PL-10s deployed in firing position. (B747SPNKG on Weibo)

Royal United Services Institute defence think tank highlighted that the PL-10 had a 'superior kinematic performance to the American AIM-9X Sidewinder,' which was the most capable in the USAF's arsenal.[110] The missile was one of very few of its kind in the world capable of engaging at near beyond-visual ranges, with some reports indicating that it had a lock-on-after-launch capability and could receive targeting data by datalink during flight to facilitate engagements at greater distances.

The PL-10's high performance supplemented the advantages provided by the J-20's unique side launch rails, which deployed the missiles outside the fighter's side weapons bays with bay doors closed to conserve stealth and facilitate much-faster launches during short-range engagements. This contrasted with American and Russian fifth generation fighters which lacked such a launch system, and consequentially could not launch missiles without both significantly increasing their RCSs and disturbing airflows around them. This compromised their flight performance which could be particularly detrimental in close-range engagements. Avoiding the need to open its weapons bays, combined with the PL-10's particularly long reach for a missile of its kind, created opportunities for J-20 units to use unique new tactics including operating radar silent by using passive sensors and third-party sensor data to attack at shorter beyond-visual ranges.

Long-range missiles have increasingly since the 1980s in particular emerged as one of the most important determinants of a fighter's potency in air-to-air combat. Gaining an advantage with a new generation of missiles represented the culmination of decades-long PLA efforts, with Chinese fighters having had negligible BVR anti-aircraft capabilities when the Cold War ended until the Soviet R-27ER and R-77 missiles were acquired in the 1990s. The latter was the country's first with active radar guidance, meaning it had much prized 'fire and forget' capabilities, could be used to implement a significant range of new tactics, and allowed each fighter to engage many more targets simultaneously. China itself notably financed development of an upgrade package for the Su-27 to be able to integrate the R-77 due to the almost total lack of financing for modernisation in Russia at the time,[111] and the class was followed into service from around 2005 by the indigenous PL-12 developed with Russian support and with a 100km range. The PL-12, like the R-77, had active radar guidance and provided parity with its most capable Western contemporary the AIM-120C. The entry into service of their successors the AIM-120D and PL-15 within months of one another transformed Chinese parity in BVR air-to-air missiles into a comfortable lead. As the U.S. moved to introduce a successor around 2025 – the AIM-260 – it was widely speculated that a new Chinese missile class could also be confirmed.

3 CHINA'S FIRST STEALTH FIGHTER IN SERVICE

ROLE OF THE J-20

Following the J-20's first flight in January 2011 speculation grew considerably from a wide range of sources as to what primary role the aircraft was designed to fulfil. The fighter's size, twin engines, stealth profile, bubble canopy and extensive flight control surfaces were among the design features that strongly indicated an air superiority role similar to the F-22. A number of alternative theories nevertheless also initially prevailed, however, with the large airframe, high thrust, operational altitude and speed, long range and sizeable missile payload matching the expected characteristics of a number of other aircraft classes.

The J-20 was initially widely considered by Western analysts to have been designed as an interceptor-type aircraft optimised for very-long-range air-to-air engagements against bombers and force multiplier support assets such as airborne early warning and control (AEW&C) systems and tanker aircraft. This resembled the role of Russia's MiG-31 interceptor which was by far the largest aircraft in the world built for air-to-air combat, allowing it to deploy air-to-air missiles over three times the size of those carried by fighter-sized aircraft, as well as an exceptionally large sensor suite far heavier than F-22-sized aircraft could accommodate.[1] Such assessments of the J-20 were partly influenced by erroneous initial Western estimates its size, which placed it at a gargantuan 22–23 metres and around 30 percent heavier than the F-22.

Research director at the China Aerospace Studies Institute of the USAF Air University Rod Lee was among those to express considerable certainty that the J-20 was an interceptor designed primarily to neutralise large high-value aircraft such as AEW&Cs. 'The PLA de-emphasises the importance of attrition warfare and instead advocates a "systems destruction" approach. Killing individual adversary fighters (even in large numbers) is not as useful as killing [high-value airborne assets] and key ground targets,' Lee stated, which he believed would be reflected in how the J-20 was built and would be used.[2]

Another view widely held among Western analysts was that the J-20 was optimised for air-to-ground attacks as a fifth-generation equivalent to the USAF's massive F-111 strike platform, which again stemmed largely from erroneous estimates of its size. The editor of *Defence Technology International*, Bill Sweetman, for one, referred to it as a 'stealth F-111,'[3] with the J-20 otherwise referred to as 'a fast, long-range, stealthy, precision bomber [which] might be able to penetrate U.S. defences and destroy any airfield infrastructure that survived the missile barrage.'[4] This was in part based on the general

A taxiing J-20 in yellow primer. (万全&T汪汪T on Weibo)

A J-20 seen at the Zhuhai Airshow in 2018. (垂直风行 on Weibo)

importance the PLA attributed to being able to strike U.S. bases across the region, as well of as the value of a stealth jet able to destroy enemy infrastructure and supplies as an interdictor or target ground or naval assets as a strike fighter.

Transfers of the J-20's two demonstrator airframes to the China Flight Test Establishment at Xi'an-Yanliang in 2012 for the first time allowed satellite images to show scale comparisons next to Flanker airframes there, providing an important indication of the fighter's actual size. At approximately 20.8 metres it was in the same weight range as the F-22: heavy by the standards of fighters but significantly lighter than a stealth interceptor with oversized missiles or F-111-type strike fighter would have been. Information on the J-20's manoeuvrability, which was set to improve markedly with the integration of WS-15 engines and incorporation of thrust-vectoring technologies, further strongly indicated an air superiority role rather than an interceptor or strike fighter. The J-20's weapons payload provided a further important indication of this, as the fighters were never seen with weapons other than air-to-air missiles, while their weapons bays could not accommodate the PLAAF's oversized PL-XX missiles built to neutralise high-value aircraft, with these instead carried externally by non-stealthy J-16 fighters.

The J-20 continued to be described in Western popular media and by less-seasoned analysts as a dedicated interceptor or dedicated strike fighter into the 2020s, with cognitive dissonance and an unwillingness to admit that the F-22 had a direct, symmetric and peer-level competitor from an East Asian rival playing an important role. As leading expert on the PLAAF Rick Joe summarised:

> It is difficult to assess if foreign commentary on the J-20's role reflects a genuine consideration of evidence or arises from some underlying discomfort or disbelief that a PLA stealth fighter may be intended to compete in a generally symmetric way with the F-22 and F-35, the former of which in particular has attained something of a mythic status in defence watching circles.

Joe further highlighted regarding the PL-XX that the fact the J-20 had been 'consciously designed to a size which prevented it from carrying a missile of this size should be further instructive of its role.'[5]

Although the J-20 was developed primarily for air superiority roles, the aircraft are very likely to gain secondary BVR air-to-surface and anti-shipping capabilities. This has been common to all twenty-first century fighters other than the F-22 and is considered a leading limitation on the American jet's potential. In March 2018 the J-20's chief designer Yang Wei indicated regarding its future roles: 'Of course, it will be tasked with penetrating air defence networks, but that will not be its only mission. It definitely has multiple functions. How we will use it depends on its production and deployment scale.'[6] He reiterated that as numbers grew the fighter would be seen operating in a more diverse range of roles.[7]

At the Zhuhai Airshow that November Yang observed: 'first, the J-20 has outstanding stealth capability; second, it has outstanding long-range strike capability; third, it has outstanding capability in information warfare.'[8] The fighter was described on a flyer at the airshow as 'renowned for its dominant role of medium and long-range air combat and excellent capability in ground and marine precision strike... major operational missions include: seizing & maintaining air superiority, medium & long range fast interception, escort and deep strike.'[9] With which armaments it could do so, however, remained highly uncertain with none having been seen.

As the J-20 participated in its first confirmed combat exercises in January 2018, PLAAF officer Zhang Hao who oversaw the fighter's flight testing observed: 'J-20 will be like a needle that can penetrate and break down the enemy's air-defence network.'[10] It was reported in January 2020 that alongside air superiority, the PLAAF had been simulating land attack and maritime strike roles for the J-20.[11] A further indicator came at the August 2021 Zapad/Vzaimodeystviye Sino-Russian strategic military exercises when J-20s spearheaded a simulated air raid on enemy frontline command centres and air defence observation outposts. The state-run *Global Times* reported: 'The J-20s kicked open the door and seized air superiority by accurately attacking high-value targets on the ground, and this was the leading role the J-20s had played in the exercises as stealth fighter jets... This allowed ground troops and other non-stealth warplanes to proceed with their missions with much less risks.' It was followed by non-stealthy Russian Su-30SM fighters and Chinese JH-7, J-11, J-16 and H-6 jets – serving as a force multiplier for all of them.[12] Even preceding the first deliveries of J-20s to frontline units, chief designer Yang Wei had stated regarding the missions which would be greatest priorities for J-20 units: '"Kicking down the door" is definitely a key aspect, but just using "kick the door" has certain limitations. It will also play other roles. This is also related to batch size. It's one thing when the fleet size is small, but when the number of fighters is larger it will be another story in terms of the roles played.'[13]

The J-20's role in Zapad/Vzaimodeystviye 2021 strongly indicated the fighter not only had a strike capability, but was seen as an asset which could move in first to suppress enemy ground-based defences paving the way for non-stealth aircraft to operate more safely. This mirrored the role allocated to the F-35 in American war planning,[14] and previously allocated to the stealthy F-117 in the Gulf War, with both similarly described as 'kicking down the door' for non-stealth assets to then operate more easily.[15] In January 2022 J-20s were again seen simulating strikes and electronic attacks on ground targets in exercises, with experts cited by the *Global Times* highlighting regarding the fighter's function: 'J-20 could be used to penetrate hostile defences and seize air superiority, and other types of non-stealth aircraft could be used to unload a large amount of munitions on targets.'[16] With a rapid increase in deliveries in 2022–2023 confirming that the J-20 would form the backbone of the fleet, rather than a few niche units as the F-22 had, the possibility of it being fielded as anything other than a versatile multirole aircraft further diminished, with a focus on other roles expected to grow as the fleet expanded.

Among the weapons expected to be integrated onto the J-20 is an anti-radiation missile: a weapons class well optimised for air defence suppression and aforementioned 'kick down the door' operations which operates by homing in on enemy radar emissions. With the LD-10 anti-radiation missile having been developed based on the PL-12's airframe, a successor for the J-20 is likely to be based on the PL-15's. A cruise missile with a small form factor similar to the Su-57's Kh-59MK2 and the F-35's Joint Strike Missile is also likely to be fielded. With the Air Force deploying J-20s extensively for maritime patrols particularly from 2021, and the Navy potentially eventually inducting the class, an anti-ship missile is also likely to enter service.

THE J-20 MATURES IN SERVICE: TIMELINE

On 1 November 2016 J-20 fighters in PLAAF service were first revealed to the public at the Zhuhai Airshow when a pair conducted a brief flypast at the opening ceremony. The fighters had been delivered to the Air Force some months before but were not yet considered active or operational.

In November 2017, eight months after their formal entry into service, the J-20 participated in the year's Red Sword military exercises at the Dingxin Flight Test and Training Base alongside H-6K bombers, KJ-2000 and KJ-500 AEW&C systems, JH-7 strike fighters and J-10 and J-11 fighters. Its participation was confirmed three months later when state media described J-20s as having 'played an important role in the "Red Sword 2017" systemic confrontation drill, laying the foundation for the improvement in the air force's new combat capabilities.'[17]

In January 2018 the PLAAF conducted a series of large-scale live-fire exercises coordinated across multiple airbases which put

JH-7 strike fighters (front) and J-11B fighters (behind) at Red Sword 2017 at the Dingxin flight test and training base. (China Military Online/Fan Yishu)

A closeup image of a J-20A with J-16 in background. (空天砺剑 on Weibo)

the capabilities of the J-20 and J-16 to the test. Their announcement represented the first public confirmation of J-20s being involved in any kind of exercises. J-20 pilot Chen Liu compared the task of those flying the new stealth fighter over the past several months to 'white hat' hackers and video game testers, with their job being the identification of mechanical and software glitches for fixing or debugged before the next mission or drill to break the aircraft into service. Pilots were also tasked with compiling manuals for other PLA units. It was strongly publicised that Chen's grandfather was the renowned Air Marshal Liu Yuti, a MiG-15 pilot credited with shooting down eight American fighters during the Korean War, which tied the new air superiority fighter with the historical memory of the PLAAF's finest hours guarding North Korean skies.[18]

On 9 February 2018 the PLAAF announced that the J-20 had reached a combat-ready status, with spokesman Shen Jinke highlighting that the fighter would enhance the service's comprehensive combat capabilities and its ability to safeguard China's sovereignty, security and territorial integrity.[19] Regarding the speed with which this had been achieved just 11 months after the fighter entered service, senior researcher with the PLAAF Command College Wang Mingzhi indicated that this demonstrated how the Air Force 'transformed advanced technology into combat strength faster than the air forces of other powerful nations.' Chinese defence experts widely observed at the time that the fighter was 'now ready years ahead of when the United States was expecting them to be' which would serve to 'deter the U.S.'[20]

On 9 May 2018 PLAAF spokesman Shen Jinke announced that J-20 units had begun maritime training, stating: 'The J-20 has conducted a combat training mission in sea areas for the first time, and this has further strengthened the comprehensive combat capability of the PLA Air Force... It will help the Air Force to better execute its sacred duty to safeguard national sovereignty, security and territorial integrity.' He added that pilot training for the J-20 had been proceeding smoothly.[21] Shen subsequently announced on 12 May that since the J-20 was commissioned it had participated in realistic air battle training with other fighters like the J-16 and J-10C and played an important role in confrontational drills.[22]

The beginning of maritime training was announced just days after the PLA reinforced disputed island outposts in the South China Sea with YJ-12 anti-ship missiles, HQ-9B air defence systems and advanced electronic warfare assets, and highlighted the important role that the J-20 could play in supporting broader Chinese maritime anti-access area denial capabilities.[23] PLA Navy J-11BH fighters, the J-20's direct predecessor, had been deployed to disputed islands since 2015.[24]

In the third week of May 2018, amid high tensions in the Taiwan Strait, a program on conflict over Taiwan aired by the state-run China Central Television network highlighted that the J-20 could be deployed for operations in the strait in the near future. Military scholar with the China National Defence University Wang Mingliang highlighted that the government in Taipei feared 'precision strikes on the leadership or key targets' by J-20s which 'could come and go at will above Taiwan' as adversaries would struggle to threaten the aircraft .[25] Analysts at the time questioned whether J-20s would be deployed near Taiwan for extended periods, highlighting that the fighter would likely be held back to deny potential adversaries opportunities to study it. The precedent set by American F-22 operations over Syria were cited, with USAF Lieutenant General VeraLinn Jamieson confirming that adversaries gained 'a treasure trove' of information on how the stealth jets operated due to extended periods in its proximity.[26]

A J-20 leads a J-10C and a J-16 in formation. (空军知识 on Weibo)

On 2 June 2018 the *People's Daily* Chinese state media outlet gave insight into the J-20's role when operating alongside the PLAAF's two other new fighter classes the J-16 and the J-10C in joint nighttime exercises. 'During the exercise, the J-20 pilots used their aircraft's advantages in terms of situational awareness and stealth to gain air superiority, while the J-16 and J-10C fighters carried out precision strikes on ground targets. The training sought to take full advantage of the different capabilities of the fighter jets,' the paper elaborated, with pilot He Xing reiterating that the exercises capitalised on the three aircraft's complementary attributes.[27] The role assigned to the J-20 served as further confirmation that it was indeed an air superiority fighter.

On 11 November 2018 a J-20 at the Zhuhai Airshow flew with its bay doors open providing confirmation of its air-to-air missile payload.[28] The aircraft performed significantly longer and more aggressive flight displays than at Zhuhai 2016.

In January 2020 the J-20s under the 9th Air Brigade conducted joint exercises with J-10Cs and J-16s, with the three widely referred to by Chinese sources as the 'Three Sky Musketeers' when operating together. The fighters demonstrated a new formation dubbed the 'battle box,' where formations of two J-20s, two J-16s and one J-10C worked in unison, with the J-20s tasked with launching surprise attacks on enemy aircraft while the J-10C provided cover at close ranges. This capitalised on the former's stealth capabilities and the latter's high manoeuvrability. The J-16 was then tasked with baiting the enemy. Another scenario saw J-20s seek and destroy 'strategic nodes of the enemy's defence,' in particular AEW&C systems, while the J-16 struck mobile radar installations and the J-10C gained air superiority over enemy fighters. The exercises demonstrated the growing synergy between the three classes which were all entering service in fast-growing numbers. All the three were separately produced at much higher annual rates than any other fighter classes in the world, with the exception of the F-35, which allowed the PLAAF to build a complementary combination of fighters in ways no other service could.

In July 2020 mainland and Hong Kong media outlets widely reported that a new J-20 variant, the J-20B, had entered production. The fighter was reported to have been unveiled on 8 July at a ceremony attended by senior military leaders including the Vice Chairman of the Central Military Commission General Zhang Youxia, although beyond the addition of thrust-vectoring engines none of its new capabilities were revealed. Some of the most widely speculated new features included missile racks for a greater internal payload, more powerful sensors and electronic warfare systems, and further RCS reductions. It was later assessed that this likely referred to the beginning of production of flight prototypes rather than actual serial production.

First deployed by the 4th Mixed Air Brigade established in June 1950, intensive seven-day-a-week training allowed the PLAAF to field three combat-ready MiG-15 divisions by November 1951, and four more early the following year. The vast majority of Chinese aviation's most formative combat experiences during the Cold War were gained flying the MiG-15 and its close derivative the MiG-17. The MiG-15 marked a major milestone not only for the Chinese fleet, but also for the Soviet aviation industry as its first combat jet considered a world leader in its air-to-air performance. A total of 2,375 MiG-15s are estimated to have been acquired by China. (Artwork by Tom Cooper)

Built with a similar swept wing design to the MiG-15, the F-86 much like its Soviet rival first flew in 1947, entered service two years later, and saw its first combat over Korea. Despite Chinese airmen's limited experience and training, it was quickly widely recognised that the F-86 was the only Western fighter class capable of seriously challenging their MiG-15s, leading it to be rushed into production on a larger scale. This illustration shows an F-86F of the USAF's 334th Fighter Interceptor Squadron, one of the first squadrons to field the F-86, transitioning shortly after it moved to Langley Air Force Base, before redeploying to Kimpo Air Base in Korea from November 1950. (Artwork by Tom Cooper)

Four E-3s from the 552nd are usually assigned to the Pacific Air Forces and deployed at Kadena Air Force Base, Okinawa. Having entered service from 1977, the E-3 as the U.S. Air Force's only AEW&C system is several decades older than equivalent Chinese systems, with the growing performance discrepancy particularly after the introduction of the Chinese KJ-500 having placed American and allied forces at a significant disadvantage in the Western Pacific. The superior situational awareness provided by AEW&Cs is considered particularly valuable when engaging adversaries fielding large numbers of stealth aircraft, with the decision to belatedly order a successor to the E-3, the E-7, announced in late 2022 following signs of growing concern in U.S. Air Force evaluations of the J-20 and a significant expansion in the fighter's production rates. The example shown here is an E-3G of the U.S. Air Force 552nd Airborne Warning and Control Wing. (Artwork by Tom Cooper)

The PLAAF acquired its first fourth generation fighters in December 1991 with the delivery of the first Su-27 Flankers. The Flanker in the 1990s came to be widely considered to be the world's prime air superiority fighter, and was the last Soviet or Russian aircraft ever to attain this status, closing the race between top Soviet-American fighter units where the MiG-15 had first started it. Much like the MiG-15 in the 1950s, there were significant difficulties adapting to the Su-27 due to its level of sophistication compared to previous Chinese operated aircraft, with mass acquisitions similarly transforming previously very limited PLAAF fighter capabilities into those of a leading air power. Su-27s produced under license in China entered service from 1998 as the J-11. This illustration shows a JJ-11A. (Artwork by Tom Cooper)

After China lost access to new generations of Soviet combat aircraft in the early 1960s, the F-15 Eagle's introduction into the American and Japanese fleets from 1975 and 1981 further increased the already significant performance discrepancy between their top fighter units and those in the PLAAF. The Eagle would continue to serve as America's primary heavyweight air superiority fighter for over half a century, and after playing a central role in Operation Desert Storm, where it was responsible for over 92 percent of confirmed American air-to-air kills, its demonstrated capabilities contributed to the PLAAF's perceived need for a major upgrade to its fighter fleet. This illustration shows an F-15C of the U.S. Air Force 48th Fighter Wing. (Artwork by Tom Cooper)

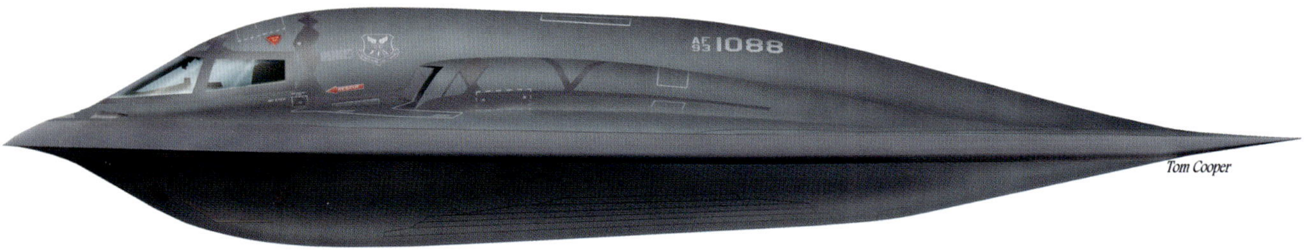

The oldest class of stealth aircraft still in service, the B-2's first combat operations were launched over Yugoslavia from 24 March 1999, which required 30-hour flights from Whiteman Air Force Base, Missouri. The concerns these operations raised among potential future American adversaries were made all the more significant when a B-2 on 7 May launched a precision strike on the office of the military attaché in China's Belgrade embassy. The incident highlighted the importance for China of fielding more modern air defence capabilities, with indigenous fighter programs playing a central role in modernisation efforts. A successor to the B-2, the B-21, first flew in November 2023, and is expected to face a near-peer Chinese counterpart being developed under the H-20 program. This illustration shows a B-2 of the U.S. Air Force's 509th Bomb Wing. (Artwork by Tom Cooper)

Developing AEW&Cs had been a leading priority for the PLAAF from the early 1990s, with the KJ-2000 being the first to enter service when it joined the fleet in 2004. Five were built, all of which deployed under the 26th Special Missions Aircraft Division, 77th Regiment. The program represented a starting point for development of AEW&C capabilities, and was followed by a succession of increasingly capable platforms, culminating in the KJ-500 in the mid-2010s which was produced on a much larger scale. (Artwork by Tom Cooper)

The KJ-500's service entry in 2018 closely coincided with that of the J-20, and provided a highly complementary asset and key force multiplier for new stealth fighter units. Approximately 35 of the AEW&C platforms were estimated to be in service by 2024, with production reported to be increasing significantly after 2021 much as was the case for the J-20. The KJ-500 was specifically singled out by U.S. Air Force officials when assessing the J-20's combat potential, highlighting the impressive command and control and missile guidance capabilities provided. This illustration shows a KJ-500 of the Chinese People's Liberation Army Air Force, 26th Special Missions Aircraft Division, 76th Regiment. (Artwork by Tom Cooper)

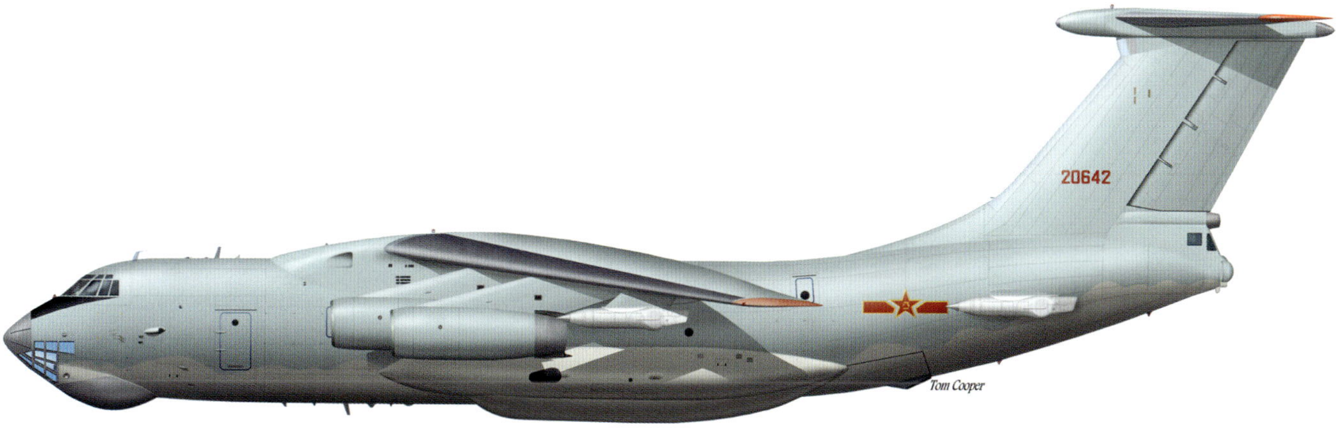

Following large-scale acquisitions of ex-Soviet Il-76 transports, China acquired three of the Il-76's aerial refuelling tanker derivative the Il-78 from Ukraine which were delivered from 2014. Commonality with the Il-76 allowed them to be easily integrated into the fleet. Mirroring the Il-76's role as a stopgap until China developed an indigenous higher capacity airlifter, the Y-20, which entered service from 2016, the Il-78 similarly provided critical experience with a large and high endurance tanker until a Y-20 derived tanker, the YY-20, entered service in 2022. The J-20 integrates a retractable in-flight refuelling probe under its fuselage to conserve stealth, and although its much greater range than Western fighters significantly reduces its reliance on aerial refuelling, availability of aerial tankers can significantly widen its already very broad area of operations to Guam, the Arctic, and the Indian Ocean. This illustration shows an Il-78 of the Chinese People's Liberation Army Air Force, 13th Transport Division, 38th Air Regiment. (Artwork by Tom Cooper)

The PL-XX is estimated to be the world's longest ranged air-to-air missile, and as it is too large for the J-20's internal weapons bays it is primarily deployed by J-16s. The missiles allow J-16s to provide critical support to J-20 units by eliminating support aircraft such as AEW&Cs and aerial tankers. Western fighters' very heavy reliance on aerial refuelling to operate in East Asia allows the J-16 to potentially seriously complicate Western fleets' ability to wage war in the region. This illustration shows a J-16, with PL-XX and PL-10 missiles, of the Chinese People's Liberation Army Air Force, 172nd Air Brigade. (Artwork by Tom Cooper)

The J-16 was China's first fighter integrating avionics and weaponry with similar fifth generation levels of sophistication to the J-20, including being the first to carry an AESA radar and PL-10 and PL-15 missiles. Although entering service only an estimated two to three years before the J-20, the aircraft was significantly more straightforward to bring to full operational readiness and into large-scale production, which made its role in China's defence particularly critical until the J-20 program further matured in the early-mid 2020s. J-16s and J-20s have consistently operated very closely together, with the first two air brigades to operate J-20s, the 176th and 172nd, both receiving J-16s shortly beforehand, while no other fighter has been reported as frequently to be exercising alongside the J-20. This illustration shows a J-16 with PL-10 and PL-12 missiles, KD-88 PGM training round with guidance pod and electronic warfare pod, of the Chinese People's Liberation Army Air Force, 40th Air Brigade. (Artwork by Tom Cooper)

Entering service from 2021, the J-16D provided the PLAAF with an electronic warfare aircraft benefiting from commonality with the service's very large J-16 fighter fleet. The aircraft has antennas and conformal electronic warfare arrays built into its fuselage and integrates jamming pods to analyse radar frequencies and locate radar transmitting devices, making it optimal both for air defence suppression, where J-20s are also expected to play a central role, and to provide electronic warfare support to fighter units. The J-16D's only overseas equivalent the EA-18G Growler is fielded exclusively by the U.S. Navy and Royal Australian Air Force, although the Chinese aircraft's capacity for carriage of sensors, electronic warfare systems and weaponry are very significantly greater, as is its range which is critical to facilitating operations alongside J-20 units. This illustration shows a J-16D, with three electronic warfare pods and a PL-15 missile, of the Chinese People's Liberation Army Air Force, 26th Air Brigade. (Artwork by Tom Cooper)

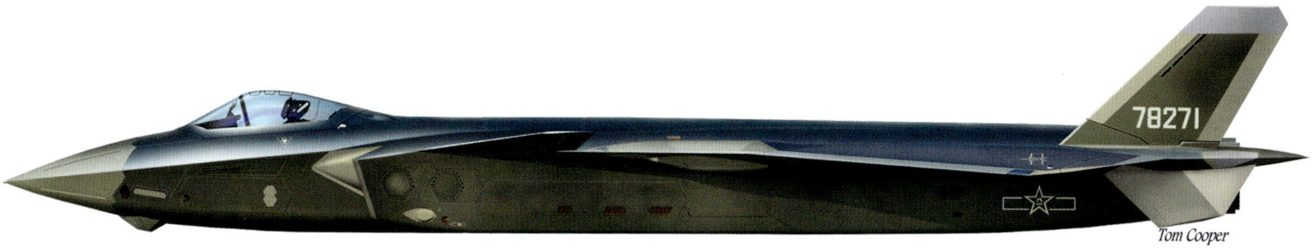

Based in the Gobi Desert, Gansu Province, the 176th Brigade became the first PLAAF unit to receive J-20s in mid-late 2016, and was assigned to develop new tactics for new fighters as well as the new J-16 and J-10C. The base at Dingxin was China's only large-scale comprehensive test training base for aircraft and air defence systems, with a comparable role to Nellis Air Force Base in the United States. This illustration shows J-20 serial number 78271 of the Chinese People's Liberation Army Air Force 176th Air Brigade (Dingxin Flight Test & Training Base). (Artwork by Tom Cooper)

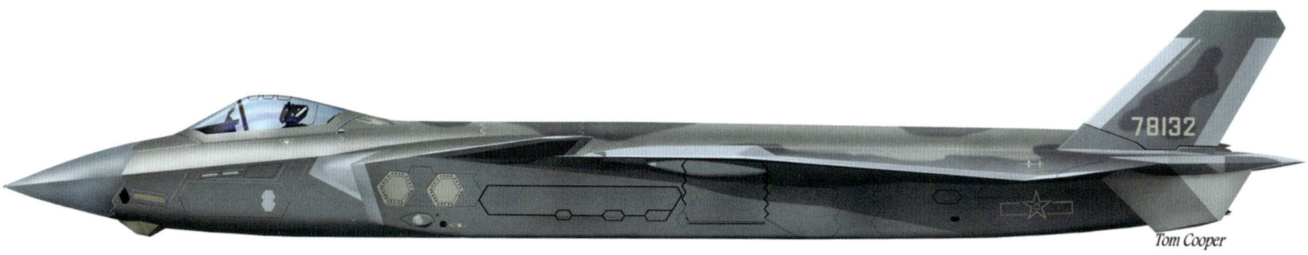

Based at the Cangzhou Flight Training Base in Hebei province close to Beijing, the 172nd Air Brigade in February 2018 became the second unit to integrate the J-20. It served as both an advanced training and a combat reserve unit. While optimally located for air defence duties over the capital, the base at Cangzhou was also one of the closest to the Korean Peninsula which in the months preceding delivery had been a leading hotspot of geopolitical and military tensions. The facility was also optimally located for operations over the East China Sea. This illustration shows the aircraft carrying serial number 78132. (Artwork by Tom Cooper)

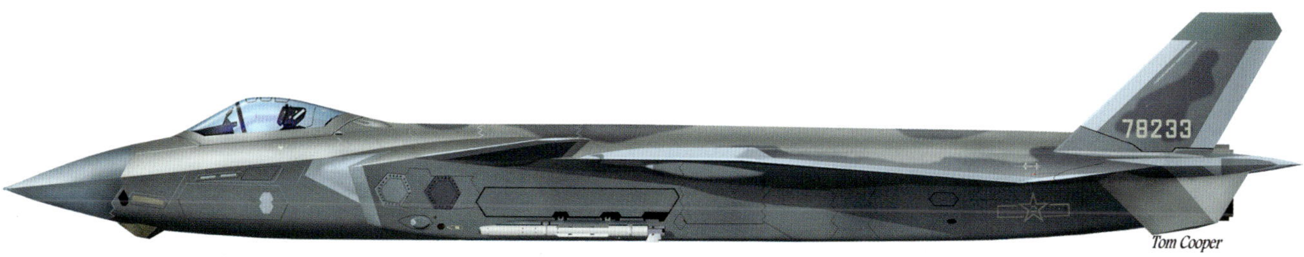

Each J-20 fighter carries one PL-10 short range air-to-air missile in each of its side weapons bays, with weapons integration tests for an entirely unique retractable missile launch rail for these bays having begun no later than the first quarter of 2013. Rails deployed the missiles outside of the aircraft's weapons bays with bay doors closed, which facilitated much faster launches during short range engagements, while avoiding the air flow disruptions and radar cross section increases caused by opening bays during combat. This illustration shows a J-20, with PL-10 missile in firing position, of the Chinese People's Liberation Army Air Force 172nd Air Brigade (Cangzhou Flight Test & Training Base). (Artwork by Tom Cooper)

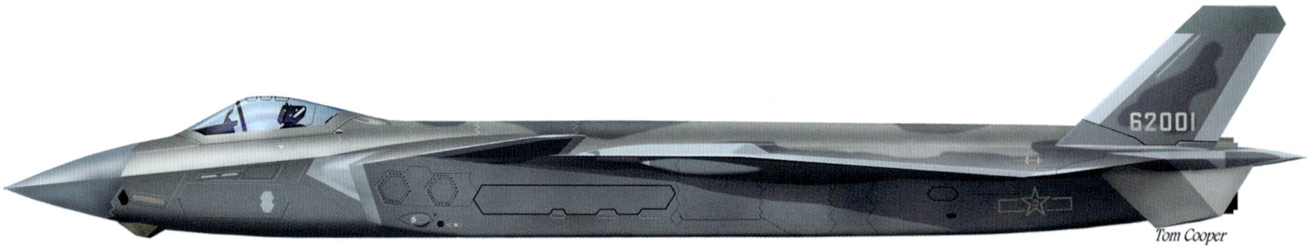

Otherwise known as the Wang Hai Air Group 'Ace Unit,' named after Korean War ace Wang Hai, the 9th Brigade is based less than 300 kilometres from Shanghai and optimally located for operations in the South China Sea, the Taiwan Strait and the Korean Peninsula, as well as for providing air defence to key southeastern population centres. The brigade had been the first in the PLAAF to receive Su-27s from 1991, then the first to field Su-30s which it transitioned to in 2000, before in January 2019 becoming the first frontline unit to field the J-20. This illustration shows J-20 serial number 62001 of the Chinese People's Liberation Army Air Force 9th Air Brigade (Wuhu Air Base, Central Theatre Command). (Artwork by Tom Cooper)

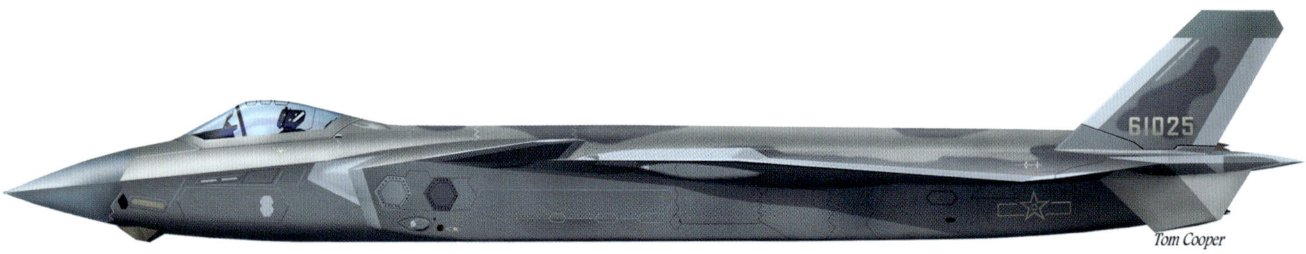

The 1st Air Brigade at Anshan in January 2021 became the first frontline unit to receive the new J-20A variant powered by indigenous WS-10C engines, marking the beginning of a new phase in the program after which deliveries to new units would accelerate considerably. Anshan Air Base represented a priority location optimal for air defence duties along China's east coast, as well as operations around Korea and across the East China Sea. This illustration shows J-20A serial number 61025 of the Chinese People's Liberation Army Air Force 1st Air Brigade (Anshan Air Base, Northern Theatre Command). (Artwork by Tom Cooper)

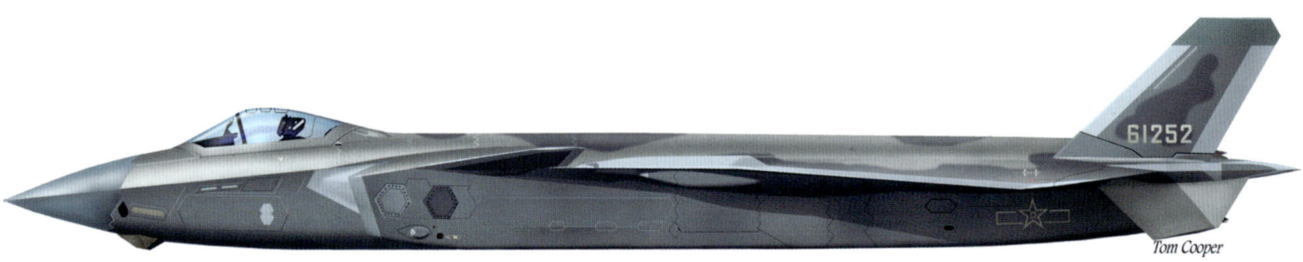

The 4th Air Brigade was confirmed in December 2023 to have received the J-20A to replace its older Su-27 and J-11 fighters, coinciding with new deployments of unmanned aircraft to the facility. The base at Foshan was optimally located for patrols of the South China Sea, operations in the Taiwan Strait, and air defence duties over key southeastern economic centres such as Shenzhen and Hong Kong and strategic naval facilities on Hainan Island. The deployment followed a major increase in J-20A deliveries in 2022, which further rose in 2023, to far exceed those of any other fighter class going to any other service worldwide. This illustration shows J-20A serial number 61252 of the Chinese People's Liberation Army Air Force 4th Air Brigade (Foshan Air Base, Southern Theatre Command). (Artwork by Tom Cooper)

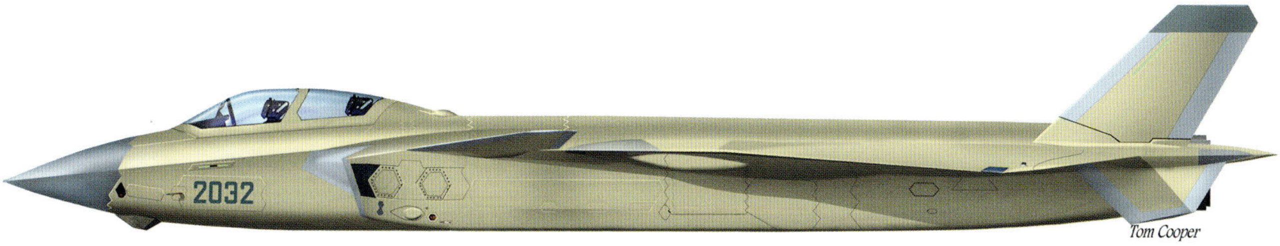

The J-20's first twin seater prototype first flew on 5 November 2021, which was also the world's first fifth generation fighter with twin seats. The aircraft are expected to serve as frontline combat aircraft rather than trainers, either alongside single seat J-20s or in their own separate units. Twin seaters provide options for better workload sharing, including by accommodating backseat weapons systems officers to operate air to ground ordinance, and are also better optimised to serving as an airborne command posts. The twin seater is expected to form the basis for development of a number of specialist variants in future. This illustration shows the J-20 twin seat prototype no. 2032. (Artwork by Tom Cooper)

Unveiled in June 2018, the Dark Sword program was seen to have the potential to provide the first unmanned aircraft that could autonomously operate in all the capacities of a fighter, making it both a potential complement to the J-20 and eventually a potential challenge to the position of manned combat aircraft in the PLAAF. It was the most high-profile of multiple unmanned aircraft that were expected to serve as 'wingmen' to the J-20, with a semi-autonomous operational capability for such a role expected to be achieved much earlier than full operational autonomy. (Artwork by Luca Canossa)

Upon entering service in 2004 the J-10 was the first class of fighter jet produced in China that was not based on a Soviet design. The program had a transitional role in catalysing the development of China's modern military aviation research, development and acquisition process, while also representing the beginning of the rise of the Chengdu Aircraft Design Institute. The J-10C which entered service in 2018 used a new high composite airframe and brought avionics and weaponry up to comparable technological level to the J-20, with production of the relatively low cost much lighter F-16-sized aircraft allowing units across the Air Force's theatre commands to be brought up to a cutting-edge technological level at a relatively low cost. This illustration shows a J-10C, with PL-15 and PL-10 missiles and three external fuel tanks, of the Chinese People's Liberation Army Air Force 2nd Air Brigade. (Artwork by Tom Cooper)

Russia's first fifth generation fighter regiment was fully equipped only in 2024, a full 23 years after the Soviet Air Force had been scheduled to field its first such regiment with fighters developed under the MiG 1.42 program. The Su-57's design differed drastically from its Soviet predecessor, and from American and Chinese stealth fighters, with a lower emphasis on stealth and a primary role of countering enemy stealth aircraft, supporting ground-based air defences, and launching strikes from in or near Russian territory using long range missiles rather than with deep penetration of enemy airspace. In contrast to the preceding Su-27, or the MiG 1.42, the Su-57 was not a serious contender for the title of a prime air superiority fighter, but was well optimised for Russia's asymmetric post-Soviet approach to aerial warfare. This illustration shows a Su-57 of the Russian Air Force's 23rd Fighter Aviation Regiment. (Artwork by Tom Cooper)

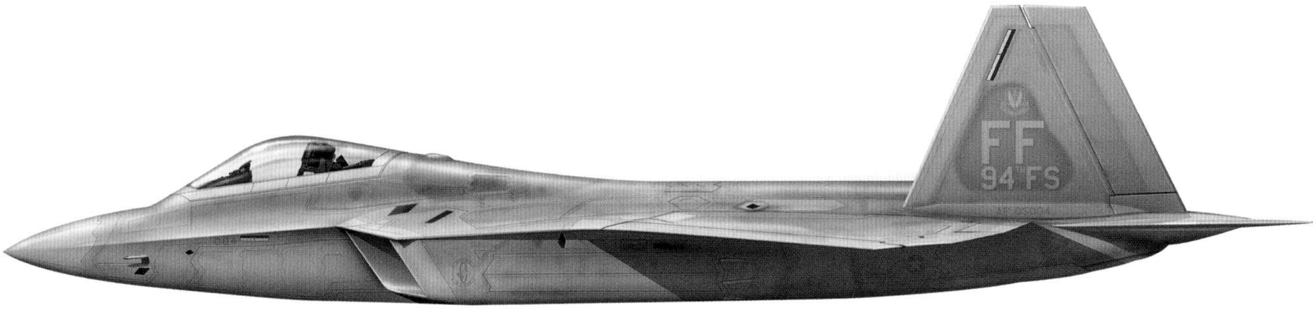

As the second squadron to receive F-22s, inducting them into service in March 2006, the 94th Fighter Squadron has since deployed the aircraft abroad multiple times from its base at Joint Base Langley-Eustis, Virginia including months-long deployments in Japan. The F-22 was initially envisaged as a successor to the F-15 as the Air Force's prime air superiority fighter, although significant issues with the program resulted not only in cuts to over three quarters of planned production numbers, but also to extremely low availability rates and limited modernisation of avionics. These and multiple other issues have led to a planned retirement of the F-22 decades ahead of initial plans, and significantly increased the burden on the F-35 to counter new generations of Chinese fighters. This illustration shows an F-22A of the U.S. Air Force 94th Fighter Squadron. (Artwork by Tom Cooper)

The program to develop a successor to the F-15 initially required a fighter with twice the combat radius on internal fuel, double the sortie rate, and half the maintenance requirements. The F-22 failed to even match its predecessor's performance, however, and not only is it estimated to be the most high-maintenance fighter class in the world, but it is also by far the shortest ranged in the world for a fighter of its size. With less than half the combat radius of the J-20 on internal fuel, the aircraft is very heavily dependant on external fuel tanks which seriously reduce its stealth capabilities. In the Pacific in particular, the F-22's short range increases its vulnerability to destruction of airbases or aerial tankers. This illustration shows a F-22A, with external fuel tanks, of the U.S. Air Force 525th Fighter Squadron. (Artwork by Tom Cooper)

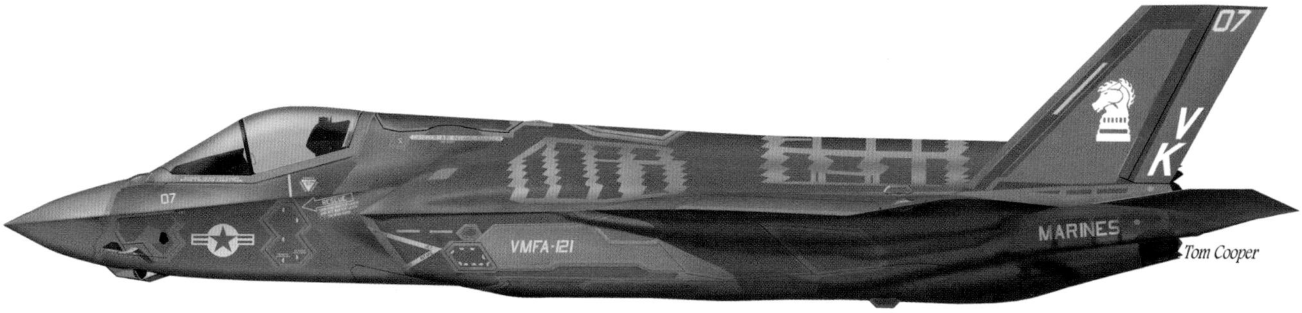

The U.S. Marine Corps was the first service to operate F-35s, and has particularly concentrated its deployments and training on preparations for possible conflict in the Pacific. The F-35B has a significantly shorter range and lower manoeuvrability than any other Western or Chinese fighter in production, and has only two thirds the internal missile payload of the U.S. Air Force's F-35A. The fighter's ability to deploy from makeshift airstrips, however, has placed it at the core of the Marines' 'island hopping' strategy for a potential conflict with China, which significantly reduces vulnerability to destruction of major airfields. The example illustrated here is an F-35B of the U.S. Marine Corps Marine Fighter Attack Squadron 121. (Artwork by Tom Cooper)

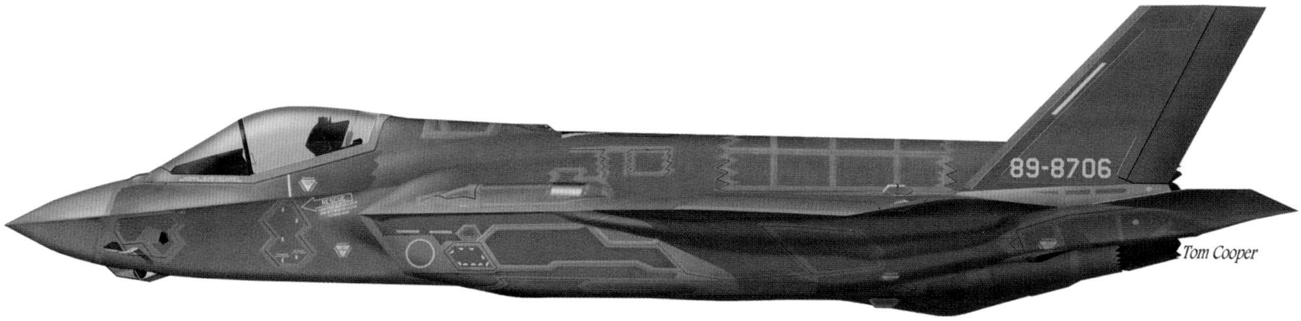

The Japanese government announced its intention to acquire 42 F-35As in December 2011, eleven months after the J-20's first flight. In 2018, the year after the J-20 entered service, Japan expanded its order to 147 F-35s, making it by far the largest foreign client for the aircraft. Ten months after the J-20 entered service, the Japanese Air Self Defence Force 3rd Air Wing at Misawa Air Base in January 2018 received its first F-35As (as shown in this illustration). The base's location in northern Japan placed it in proximity to J-20s at the 172nd Air Brigade at Cangzhou and the 1st Air Brigade based at Anshan. (Artwork by Tom Cooper)

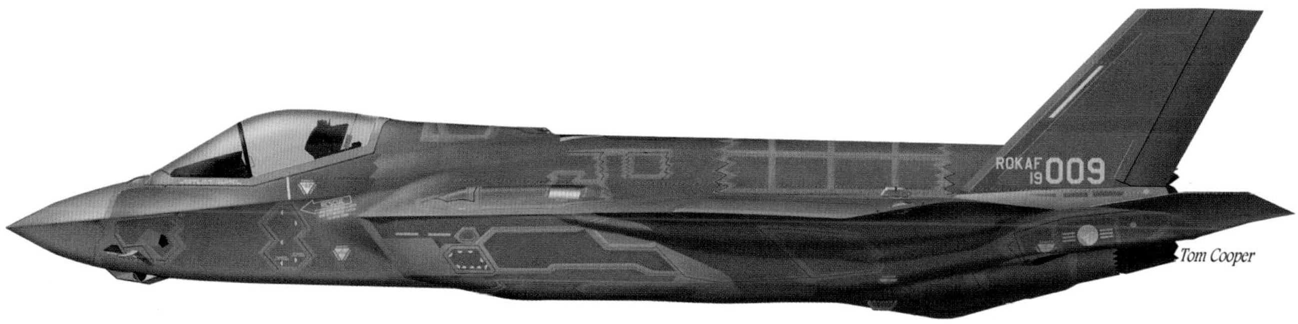

The first order from the Republic of Korea for 40 F-35As was signed in December 2014, with a follow-up order for 20 more placed in December 2023. The F-35's greater perceived ability to penetrate North Korean air defences was central to officials' arguments for acquiring the aircraft over the longer ranged but non-stealthy F-15, and unlike Japan, South Korea maintained relatively positive ties with China when initial orders were placed. Nevertheless, a worsening of Sino-Korean ties since then, Chinese treaty commitments to defend North Korea, and Seoul's obligations as a U.S. treaty ally, have given South Korea's F-35 fleet growing relevance to broader regional security dynamics. This illustration shows an F-35A of the Republic of Korea Air Force 17th Fighter Wing. (Artwork by Tom Cooper)

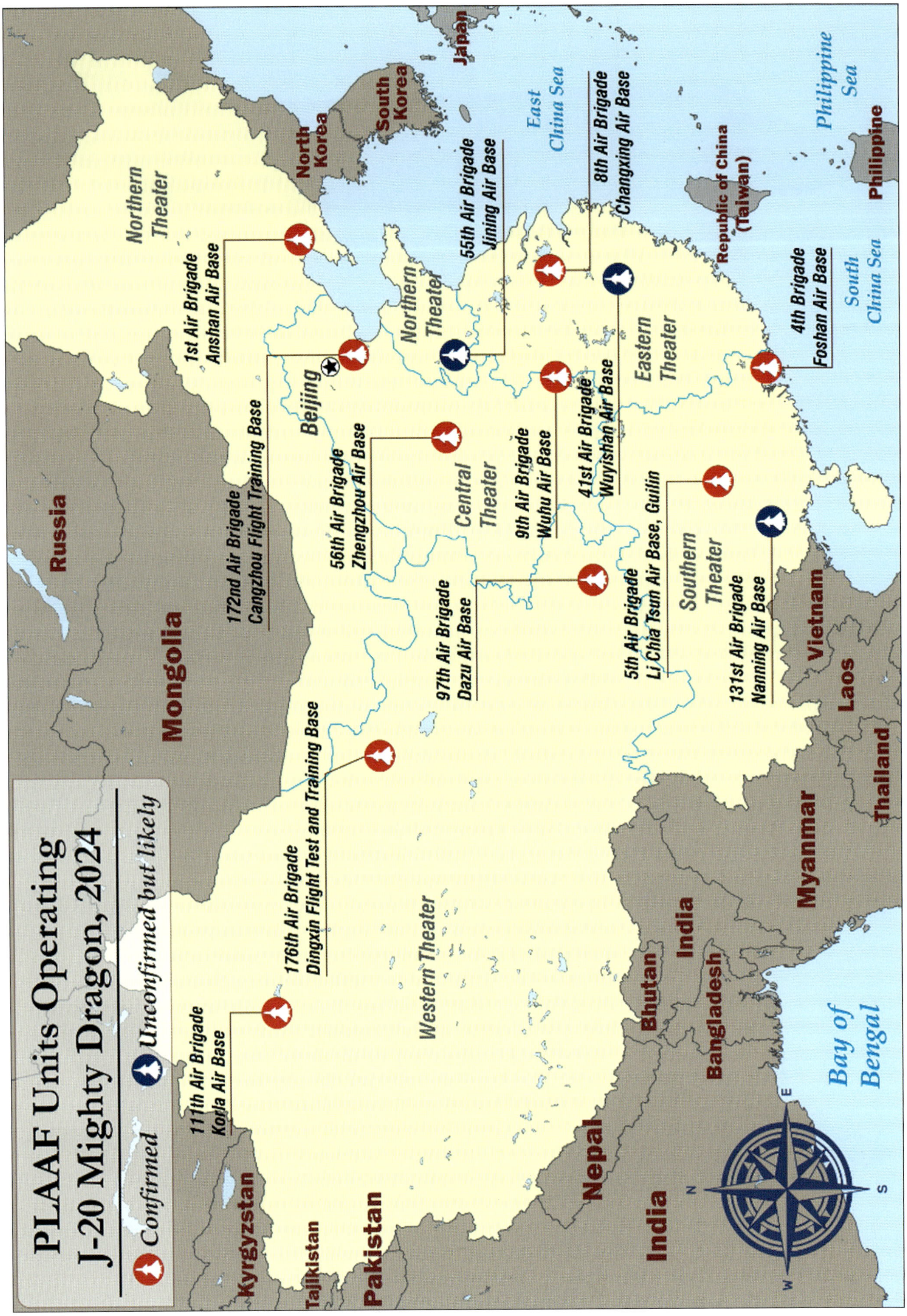

(Map by Tom Cooper)

Water vapour forms around a People's Liberation Army Air Force J-16. (XMN-5380 on Weibo)

In August 2020 two J-20s were reportedly sighted at Hotan Airbase, around 320km from the disputed Ladakh region on the Sino-Indian border, with the location providing the high-endurance fighters with coverage of most of Indian airspace.[29] This occurred just two months after border skirmishes between the two countries. Aside from deployments to the Dingxin Flight Test & Training Base in the Gobi Desert, which was much further from India and primarily for test and training purposes, this was the first known J-20 deployment under the Western Theatre Command.

In September 2020 a novice J-20 pilot confronting targets in mock air-to-air combat reportedly scored 17 kills for zero losses. This was based on a publication from the official *PLA Daily*, which although not naming the fighter which had achieved the simulated kills confirmed the 9th Air Brigade had been responsible: a unit which deployed only J-20s.[30] Such a kill ratio was highly plausible for the J-20, particularly against older PLAAF fighters such as the J-11A which were technologically around 30 years behind. The fact that this was achieved by a pilot with only 100 hours of experience in the aircraft demonstrated the growing degree to which technology could compensate for skill or experience, mirroring reports from the USAF of novice pilots seven or eight flights out of training in F-35As comfortably outperforming veteran flyers in older fourth generation aircraft.[31] Such tests were important for shaping the future acquisitions priorities of both air forces.

On 4 April 2021 at the Qingming Festival, the traditional day of tomb-sweeping, Central China Television (CCTV) gave considerable publicity to J-20 pilots paying respect to their predecessors who fought in the Korean War. Coverage emphasised that the Wang Hai Air Group 'Ace Unit' (9th Air Brigade) of heroic pilot Sun Shenglu had been equipped with J-20s, again drawing strong links between Chinese combat aviation's finest hours and the J-20 program. 'Air Force pilots in the new age will inherit the spirit of "aerial bayonet fighting," train to prepare for combat, be ready at all times for combat, and resolutely safeguard national sovereignty and dignity,' J-20 pilot Sun Teng said at the time.[32] J-20s were seen exercising in 'stealth mode' without Luneburg reflective lenses,[33] which Chinese state media sources highlighted indicated a higher state of combat readiness had been attained.[34]

In January 2022 the *Global Times* reported that the J-20s under the Northern Theatre Command, presumably the 1st Air Brigade, 'started the year with intensive mock combat training sessions' pitting them against the J-11B, J-16 and other PLAAF fighter classes. The J-20s were reportedly handicapped with Luneburg lenses which nullified their stealth capabilities, although they prevailed using performance advantages other than stealth. Both the J-20s and their adversaries were able to evade multiple simulated missile attacks by pulling high-g manoeuvres after at one point firing simultaneously on one another. Some two-versus-two mock combat lasted more than an hour, while other exercises involved live-fire attacks against groups of targets.

J-20s conducted some nighttime exercises without Luneburg lenses, with brigade commander Senior Colonel Li Ling elaborating: 'During air combat at night, the J-20 takes advantage of its stealth capability, carries out beyond-visual-range combat missions and plays the role of a commander in the air, taking initiatives in combat.' At least one mock engagement saw J-20s engage J-16s two-on-two, with the J-16 considered the PLAAF's second most capable fighter class overall. Experts cited by the *Global Times* highlighted that these exercises would give pre-fifth generation units experience in tackling fifth generation targets, with lessons learned against the J-20 likely being applicable against F-35s and other stealth targets.[35]

Against the J-16, the J-20 reportedly 'took advantages of its stealth and attack capabilities and realized the tactical goals of finding the enemy first, firing missiles first, breaking away from combat first and destroying the target first,' according to state media reports. The J-20s were nevertheless spotted and targeted at range by their adversaries on at least one occasion, although they managed to evade. The results of these exercises and feedback from airmen participating had the potential to influence the PLAAF's future acquisitions investments. If the J-20 proved overwhelmingly superior to the J-16, which was by far the most advanced Flanker in the world and one of the most capable non-stealth fighters, it could speed up a decision to phase the class out of production. If the J-16 was still capable of challenging the J-20 with full stealth capabilities, however, it may be more likely to remain in production longer, with this indicating that it would also have some viability against enemy fleets formed predominantly of fifth generation aircraft.[36]

The USAF had previously similarly sought to test its fifth generation F-22s against fourth generation fighters in the 2000s, with the stealth jets gaining overwhelming victories in simulated engagements against F-16s and F-15s using mechanically scanned array radars. Captain John Teichart, a Raptor pilot who flew F-15s against the F-22 in simulated combat, reported regarding his adversary's advantage: 'It's like clubbing baby seals it's so easy,' with pilot Captain Jeremy Durtsch who also participated similarly referring to the Raptors advantage as 'clubbing baby seals.'[37] Advanced Chinese fourth generation fighters like the Chinese J-16 and J-10C, however, which were equipped with AESA radars several times as powerful than those used in exercises against the Raptor, as well as infrared tracking systems, modern data links, and training in counter-stealth tactics, had a much greater chance of challenging fifth generation aircraft like the J-20 and F-22. Some limited stealth capabilities of their own and much more capable PL-15 missiles with inertial AESA radar guidance were other important capabilities the J-16 and J-10C had, but advances in radar technology and data sharing capabilities were likely the most significant in making fifth generation aircraft's advantages less extreme. Although fifth generation fighters would still retain a strong edge, these very high-end fourth generation aircraft were much more challenging to tackle in contrast to the 'baby seals' position more basic fourth generation aircraft had been in.

In March 2022 Commander of the U.S. Pacific Air Forces (PACAF) Air Force General Kenneth Wilsbach confirmed that the J-20 and F-35 had had their first ever encounter, which took place over the East China Sea.[38] This was the first ever encounter between fifth generation fighters from opposing sides, with China being the only country in the world outside the Western sphere of influence to field such aircraft at squadron level strength. Speaking to Wislbach at the time, former (PACAF) Vice Commander Lieutenant General (ret.) David Deptula emphasised that 'clearly fifth generation aircraft are going to become increasingly important in the Pacific.'[39]

In the second week of April 2022 the PLAAF was confirmed to have made its first deployments of the J-20A with domestic WS-10C engines for routine patrols over the East and South China Seas, with the *Global Times* highlighting that this had been facilitated by the integration of the new WS-10C engine. This followed the deployment of a U.S. carrier group to the region earlier that week – the first that decade – and was seen to reflect growing confidence in the J-20's capabilities. The *Times* referred to the deployment as intended to 'better safeguard China's airspace security and maritime interests,' with Chengdu spokesman Ren Yukun highlighting that patrols of the two seas were now considered 'training routine' for J-20 units.[40]

Griffith Asia Institute Visiting Fellow Peter Layton told *CNN* at the time that by launching the new patrols China's message to the world was: 'Any foreign military aircraft intruding into China's claimed airspace in the East and South China Sea may now be intercepted by J-20s.' The PLAAF 'now has in regular service a fleet of advanced stealth fighters as good as the Americans, who remain the benchmark,' he stated, highlighting that the J-20's high endurance would allow it to patrol further out to sea than other fighters and patrol Chinese claimed waters for longer periods. He added that the United States and Japan would be 'keenly collecting electronic intelligence data' on patrolling J-20s' stealth characteristics and communications.[41]

In June 2022 the PLAAF made its first confirmed deployment of J-20s to identify foreign military aircraft entering China's Air Defence Identification Zone. The incident took place over the East China Sea, and was reported two months later, with state media highlighting the value of the fighter's photoelectric aiming system and distributed aperture system, as well as its main radar, in such identification particularly against American stealth fighters based in the region.[42]

On 3 August 2022 Beijing announced unprecedented live-fire military exercises surrounding the Taiwan Strait as an immediate response to U.S. Speaker of the House Nancy Pelosi's visit to Taipei. J-20 deployments were referred to as being intended both to increase preparation for possible military solutions to the technically-ongoing Chinese Civil War across the strait, and to deter possible separatist aspirations by the ruling Democratic Progressive Party in Taipei, with their role in the exercises being central. J-20s were subsequently reported more frequently to be deployed for patrols near Taiwan.

In September 2022 J-20s were assigned to escort Y-20 airlifters repatriating the remains of Chinese personnel from South Korea who had been killed during the Korean War. The deployment of the new stealth fighters was described by the state-run *Global Times* as one which 'will not only pay tribute to the... martyrs but also display the increasing number of J-20s in service.'[43] The J-11B had previously been assigned to escort missions for the highly symbolic repatriations from Korea, with the use of the J-20 providing yet further grounds for Chinese commentators to widely remark on the fighter's importance in the context of the historical memory from the Korean conflict. J-20s were subsequently assigned to perform all future escort missions for veterans' remains.[44]

In October 2022 it was reported that J-20s conducting combat patrols over the East China Sea had been deployed to intercept and drive away foreign fighter jets. The location indicated they were most likely American or possibly Japanese, both major F-35 operators, while references to the targeted aircraft as 'advanced warplanes like the J-20 and others' potentially implying that the incident had seen the two stealth fighter classes again encounter one another.[45]

In January 2023 footage aired by CCTV showed J-20s performing an 'excellent job on regular patrols and controlling the East China Sea Air Defence Identification Zone,' and being scrambled to intercept intruding foreign fighters. Pilots from the Wang Hai Air Group (9th Air Brigade) Yang Juncheng and Wei Xin were shown being scrambled to intercept two unspecified incoming foreign aircraft, with Wei in his J-20 telling the targets in English: 'This is China Air Force. You have entered the Chinese ADIZ [Air Defence Identification Zone]. Report your nationality ID and the purpose of flight.' There was subsequently considerable speculation on Chinese social media that the footage showed the interception of American F-35s, which were among the most commonly flown foreign fighters in the area. As the squadron's commander Yang stated regarding its responsibilities: 'No matter what happens in the air, even if it means sacrifice, we will never step back from our positions.'[46] Pilot Wei Xin commented on patrols further south around the Taiwan Strait, which highlighted the very wide area of operations which the long range stealth fighters could have and the multiple hotspots they could cover from their base at Wuhu.[47] The growing complexity of the exercises J-20s were involved in that month were seen by analysts as a sign of improving operational capabilities.[48]

On 27 July 2023 the PLAAF commemorated 70 years since the conclusion of the Korean War, with Senior Colonel Zhang Hong from the Wang Hai Air Group observing regarding a displayed MiG-15 fighter: 'I can imagine how the older generation fought against the world's top air force with them. Winning such glorious results must have been hard-fought.' He added that the Wang Hai Air Group emerged from blood and fire in Korea, and that its personnel would continue to carry on this spirit. Through the air group and its namesake, state media at the time again drew the link between the memory of the Korean War and the J-20's status highlighting that the Wang Hai Air Group was the first combat unit to field the aircraft which were direct successors to the MiG-15 in PLA service.[49] The previous day the J-20 had made a prominent appearance at the opening of the Changchun Air Show, where aerobatics were later performed to commemorate Chinese successes in the Korean War.[50]

NUMBERS IN SERVICE

In October 2017 CCTV 4 reported that the J-20 was set to soon enter 'stable' mass production, which was interpreted to mean serial production at a fixed regular rate. Experts at Chengdu Aerospace at the time projected that the fleet would exceed 100 aircraft by the end

J-20A fighters in formation. (野生灰熊 on Weibo)

A J-20A the Zhuhai Airshow in 2022. (太湖啥个 on Weibo)

of 2020,[51] which when the fighters soon afterward began to enter service much faster was not considered unrealistic. Ten years after its first flight, and five years after entering production, by the beginning of 2021 four PLAAF units deployed J-20s, which alongside further numbers which had been built but had yet to join the PLAAF made production of close to 100 aircraft by January 2021 plausible. As with any new fighter initial production rates began relatively low before gradually rising.

Two main means were used to track the number of J-20s in service, one of which was observing serial numbers from pictures of new airframes to extrapolate total numbers. While there had been relatively frequent pictures taken at Chengdu Aerospace from 2011–2016, however, these became increasingly more difficult to observe after the fighter entered service, meaning analysts were often forced to revise their counts significantly after new airframes were seen at air shows or other events. The second means was to look at how quickly new units received J-20s, with the time between deliveries to new units growing shorter as production increased. With reports of expanded production made in October 2017, October 2021 and later in November 2022, each case was followed by strong indications that J-20s were indeed being delivered more quickly.[52]

A major landmark in the expansion of J-20 production was the Chengdu Aircraft Industrial Group's announcement in December 2021 that production had been increased in the fourth quarter of 2021, with advances in the program including development of the J-20A variant having influenced the decision.[53] This was referred to as an increase to 'full scale mass production.'[54] The state-run *Global Times* cited military aviation expert Fu Qianshao reporting at the time: 'In a short time, we will be able to see J-20s operated by all eastern, southern, western, northern and central theatre commands, and become the main force to safeguard China's sovereignty and territorial airspace security.'[55]

The J-20 was already entering service rapidly before the expansion of production in late 2021, and in October 2021 *Air Force Magazine* estimated that the PLAAF had over 150 of the fighters in service and that the scale of production only further increasing.[56] The magazine was published by the Virginia-based professional military and aerospace education association, Air Force Association, which was closely connected to the USAF. The *Global Times* at the time quoted the J-20's deputy designer Wang Hitao saying industry could 'satisfy any level of demand from the People's Liberation Army Air Force for the J-20,' and that for development 'particularly for equipment like the J-20, we need to do it faster in all aspects, including designing, production, testing, and crafting.'[57] Chinese experts at the time cited the high frequency of J-20 exercises as an indicator that the fleet was indeed large.[58] In November chief designer Yang Wei alluded to the existence of a larger J-20 fleet than had previously been estimated.[59]

In November 2022 new serial numbers observed at the Zhuhai Airshow showed that the J-20 fleet had reached approximately 180 airframes at the very least. Annual deliveries for 2022 were estimated at approximately 60–70 aircraft, meaning a fleet size of around 210–220 aircraft was estimated by the end of the year, although some estimates were significantly higher. While American estimates in early 2022 that there were 'some 200 J-20s' in service had been seriously questioned by many analysts,[60] the new serial numbers observed in November strengthened the consensus that the fleet was indeed much larger and deliveries had increased substantially. Satellite footage of facilities at Chengdu were cited as an indicator that production lines were expected to grow significantly further still, with some unverified reports indicating that J-10C production could be moved elsewhere to allocate more capacity to J-20 production. The PLAAF was in the early 2020s estimated to be receiving more new fighters every year than any other air force – around 14–20 percent more than the USAF was receiving – with production increases leading the service to invest more in better and faster training programs for new generations of pilots to keep up. The Shijiazhuang Flight Academy in Hebei Province pioneered many of these changes.[61]

The J-20 was from late 2021 the only fifth generation fighter in the world that could be described as in 'full scale mass production' and the first of its generation to reach such a state,[62] with the F-35 continuing to be denied Pentagon authorisation for full-scale production and never expected to reach even half of previously projected production rates.[63] The F-22, with orders cut by 75 percent, never exceeded 40 percent of the planned full scale production rate,[64] while with an output of around two fighters per year Su-57 production was still negligible at the time. From an estimated 60–70 J-20s delivered in 2022,[65] annual deliveries rose further in 2023 so that by the middle of the year it was expected that around 80 would be delivered. Both Western and Chinese sources reported that deliveries would rise to over 100 a year in 2024 and likely reach around 120 per year in 2025.[66]

4
THE J-20'S FUTURE AND ITS SUPPORTING ASSETS

PRODUCTION RUN

The number of J-20s the People's Liberation Army intends to commission has been the subject of growing speculation as the program has matured, with multiple factors contributing to the uncertainty. The first is that it remains unknown whether the PLAAF will procure another lighter fifth generation fighter for a complementary high-low combination – much like it acquired the lightweight J-10 to complement its heavyweight Flankers. Cost effectiveness calculations for the J-20 and any potential lighter counterpart, particularly pertaining to their operational costs which form the large majority of any fighter's lifetime costs, will likely be central to determining this.[1]

Heavyweights have formed a fast-growing proportion of China's fighter fleet from 1992, and the PLAAF has in the past transitioned its brigades from deploying lightweight fighters to replacements twice or three times their size despite the tremendous resulting increase in operational costs. Notable examples having included J-20s replacing J-10s in the 1st and 5th air brigades, which were under half their size and had under half their engine power, and J-16s replacing J-7s under a third their size and with under a quarter their engine power as seen in the 7th Air Brigade. It remains uncertain how far this will go into the 2020s, but the fact that J-20s have replaced not only Flankers but also J-10s, indicates that the fighter program could fuel further expansion of the heavyweight fleet as a proportion of all PLA fighter units. This would continue the trend that began with the large Su-27 and Su-30 acquisitions of the 1990s. The distances separating Chinese air units from their leading potential adversaries, the importance of being able to loiter for extended periods over the Pacific, and the benefits of being able to contribute to operations from a much wider range of dispersed air bases, have all been important factors in the favour of an increased focus on high-endurance heavyweight fighters.

Even if the proportion of heavyweights is not significantly increased from the levels of the early 2020s, the J-20 and its derivatives could still see close to 1,000 aircraft produced. The PLA by the early 2020s already fielded a larger fleet of heavyweight fighters than NATO and Russia combined with over 700 fighters from the Flanker family in service in 2022 (approximately 600 in

H-7 strike fighter flying through a valley at very low altitude during a training exercise. (China Military Online/ Yang Pan)

the Air Force and 150 in the Navy). Replacing the bulk of these, and possibly a portion of the country's more than 250 JH-7 heavyweight strike fighters as well, provides room for a very large J-20 production run without reducing the portion of lighter fighters in the fleet.

The second major cause of uncertainty is that it remains unknown what roles the J-20 will fulfil, with dedicated strike variants and AEW&C variants being among several to have been rumoured. The J-20 would inevitably see far more airframes built if it could perform a wider range of roles beyond air superiority – much like the Flanker saw very large production runs in both China and Russia as more well-balanced multirole variants, as well as specialised air defence suppression, strike, aircraft carrier, long-range interceptor and AEW&C variants, were all developed.[2] Future derivatives of the J-20 could potentially have different designations (J21, JH-20, KJ-20 etc.) as they perform different roles – much like the J-11's derivatives were redesignated J-15 and J-16 in the 2010s and would far outlast the J-11 itself in production.

The third source of uncertainty is the unknown extent to which manned fighters will play a role in a future PLA tactical fleet, with the J-20's production run potentially being heavily influenced by how quickly viable autonomous alternatives are developed. If J-20s begin to serve primarily as command-and-control assets for autonomous aircraft, and autonomous unmanned aerial vehicles (UAVs) come to be strongly preferred for actual combat as they are expected to when AI and other associated technologies advance sufficiently, then this will likely limit production considerably. U.S. Air Force assessments have projected a balance of around five unmanned combat aircraft for every manned one, and possibly eventually as many as 20 to one, meaning deep cuts to production of manned fighters such as J-20s

J-16 fighters. (@央广军事 on Weibo)

J-20s taxiing. (万全&T汪汪T on Weibo)

and F-35s in both fleets are highly possible.[3] Expectations regarding how soon advances in the capabilities of unmanned aircraft could cause deep cuts in orders to manned ones differ significantly, and although USAF Secretary Frank Kendall indicated in March 2022 that F-35 production could continue until around 2037,[4] Chinese defence circles have long projected much earlier dates than their American counterparts for autonomous unmanned systems taking on a much greater role in combat.

Beyond unmanned aircraft, the J-20's viability into the 2030s as sixth generation fighters join the PLAAF and the USAF is a further important factor. Whether it proves straightforward to modernise and able to cost effectively contribute to operations in the next-generation era will be key to determining how long production lasts.[5] A key overseas example was the American third generation F-4D/E fighter, which was seen to have limited modernisation potential and be far less cost-effective than fourth generation fighters such as the F-16 leading to it quickly being be phased out of production. Its successor the F-15, however, was still seen to be operationally cost-effective in the fifth-generation era when modernised to a '4+ generation' standard, and thus remained in production for well over twice as long.

A further unpredictable factor is whether the J-20 could eventually be offered for export, particularly as the extent of foreign demand depends largely on geopolitical factors which are difficult to forecast as well as on the performances of foreign economies that are seen to be potential clients. The F-35, for example, is expected to see its production run lengthened significantly by production exclusively for export, much like its direct predecessors the F-5 and F-16 did – the latter by over 20 years. Although currently built for domestic use only, the J-20 may be offered abroad after a more capable sixth generation fighter enters service around 2030.

In May 2018 the U.S. House of Representatives Permanent Select Committee on Intelligence was informed by senior experts that production of up to 500 J-20s was expected.[6] The speed at which production lines grew in subsequent years quickly led to projections emerging for a production run of well over 700 – placing it in same range as initial projections for the F-22 production at 750 aircraft. The rapid and significant expansion of J-20 production facilities provided an important early indicator of a very large intended production run. Pressure on China's defences from the tremendous numbers of F-35s that were set to be deployed by its potential adversaries, with close to 2,000 expected to enter service even if cuts were made, contributed to predictions of a large production run.

By 2023 J-20 deliveries were estimated to have reached approximately 80 per year, with both Western and Chinese sources reporting that output would rise to over 100 and then around 120 in 2024 and 2025.[7] This was a major landmark in Chinese aviation history, as the last fighter produced on any remotely comparable scale had been the J-6 – an early second generation design which first flew in 1952 and was under a quarter the size of the J-20. Even at estimated 2023 production levels the J-20 was already being delivered at comparable rates to American and Soviet air superiority fighters in the 1980s – a scale unseen for any heavyweight fighter class after the Cold War and matched only by the much smaller F-35. With the J-6 having formed close to three-quarters of Chinese fighter units in the 1980s the J-20's massive production scale indicated that it could be intended to similarly equip close to half, if not a majority, of all Chinese fighter units. No class since the J-6 had come close to doing this.

With the PLA Air Force being the only client, reports of production rates to be achieved in 2024–2025 meant it would be receiving the fighters at around 250 percent the rate at which the U.S. Air Force received F-35s – 48 per year – and closely rival the full sum of F-35 production for all 19 global clients of around 140 per year.[8] This was particularly notable since the F-35 was a much lighter single-engine fighter a little over two-thirds its size. F-22 annual production, initially expected to reach 60 per year, had briefly peaked only at 24 aircraft and averaged 19.4 per year,[9] while Russia had not made sustained acquisitions of any fighter at over 17 per year – despite acquiring fourth generation aircraft that were much cheaper. The increase in J-20 production numbers in mid-2023 notably closely coincided with deep cuts to F-35 deliveries that year due to a range of developmental difficulties,[10] which were followed by significantly deeper cuts in 2024.[11] With the fleet expected to exceed 500 fighters in 2025 the new delivery rates made a production run of over 1,000 J-20s appear likely – placing it in a league of its own with the F-35 not only in terms of the kinds of technologies they integrated, but also the scales on which they were produced and fielded among post-Cold War aircraft.

J-20 VARIANTS

By the end of 2022 five serial production variants of the J-20 had been reported and four had already begun flying, making it the only fighter of its generation to have multiple variants completed for air force use. The four already unveiled variants are listed in order of appearance.

J-20 twin-seater prototype no. 2032. (鼎盛风清 on Weibo)

J-20S twin-seater prototype no. 2031. (航空 EXIA on Weibo)

J-20: Initial production variant using a derivative of the Russian AL-31FM2 engine. Approximately 40 are thought to have been built. Produced at a low rate for 3.5 years from to late 2015 to early 2019.

J-20A: Produced from mid-2019 and inducted into service from January 2021. Beyond the integration of the WS-10C engine, the full extent of changes to the design remains uncertain. Deputy chief designer Gong Feng confirmed a redesigned airframe, structure, pipelines, electric circuits and subsystems, with one conspicuous difference being the lines on the fighter's side missile bays.[12]

J-20S: A twin-seat variant based on the J-20A which first flew on 5 November 2021 with no significant structural changes visible beyond the cockpit. The first leaked footage of the aircraft carrying out taxiing tests emerged on 26 October confirming it as the world's first fifth generation fighter with a twin-seat configuration. The *Global Times* referred to it as J-20S,[13] although a number of experts have referred to it as the J-20AS. These designations indicate that it is believed to be intended as a fully capable combat aircraft in its own right much like the J-10AS and J-11BS, in contrast to dedicated trainer models which receive 'JJ' designations such as the J-6's trainer variant the JJ 6. At the Zhuhai Airshow in 2021, a few days before the J-20S was first seen, chief designer Yang Wei revealed that a twin-seater 'would not be a trainer aircraft, because it would be developed for the enhancement of the aircraft.' PLAAF pilots had in the past strongly emphasised the advantages twin-seat configurations provided in combat, notably on the J-16 which had no single seat variant, with Chinese analysts similarly widely alluding to the advantages a twin-seater had for a range of combat situations as well as the wide-ranging roles it could perform such as command and control.[14]

Twin seaters could potentially form a large proportion of J-20 units, possibly mirroring how PLAAF Flanker acquisitions in the late 2010s transitioned from predominantly single seaters to exclusively acquiring twin seaters. Official artwork preceding the J-20S' unveiling showed twin seaters operating in formation with no single seaters among them, indicating a possible intention for them to form entire units for frontline combat operations like the J-16, rather than a small portion of predominantly single seat units as seen with the J-11BS.[15] Indeed, the Ordnance Industry Science Technology military magazine reported in August 2022 that the J-20S could be developed to perform a broad range of roles including AEW&C, electronic warfare, reconnaissance and command and control of unmanned aircraft. 'In recent years, experts' views on the position of fifth generation aircraft in information warfare have changed, as the aircraft can perform more duties than in the past,' it observed.[16] The twin-seater is expected to potentially form the basis of multiple new variants such as strike and electronic attack aircraft, much like Soviet and

The first image of the J-20B prototype. (我是痞子好吧)

with special roles were under development. Yang further indicated that advances in China's military aviation sector meant it no longer needed to bridge gaps with competitors, providing greater freedom to innovate and invest in new capabilities.[21] The J-20's status as the only heavyweight fighter of its generation in production on a meaningful scale, and the only one ever expected to receive such levels of funding, made a range of developments unprecedented overseas appear highly possible.

Russian interceptor, strike and AEW&C Flanker derivatives were based on the Su-27 twin-seater.[17]

J-20B: Unconfirmed reports of an enhanced J-20 variant under development as the 'J-20B' emerged from a range of mainland and Hong Kong sources from the late 2010s, with the addition of thrust-vectoring engines for enhanced manoeuvrability being common to all of them. Hong Kong media outlets reported that the aircraft's design had been shown to Vice Chairman of the Central Military Commission General Zhang Youxia, who was responsible for weapons development, on 8 July 2020.[18] Subsequently in December 2022 the first images emerged of a J-20 prototype numbered 2051 with a significantly evolved airframe design, which included a much flatter cockpit canopy with a distinctive low-profile appearance blending into the fighter's raised spine and a nose radome with slight geometric revisions. The changes appeared intended to improve its aerodynamic performance and further reduce its RCS, with the enlarged spine also expected to accommodate more fuel or possibly new avionics.

While prototype no. 2051 was powered by WS-10C engines, a second prototype numbered 2052 on 29 June 2023 became the first J-20 airframe to fly with WS-15 engines in twin configuration. First testing the new airframe design with an already widely used engine, before testing it with the next-generation powerplant, was in line with the Chinese defence sector's longstanding tendency to apply upgrades incrementally. The WS-15's integration made it almost certain that the new aircraft was the next iteration of the J-20 and a successor to the J-20A that had entered service in 2021, with further significant improvements to subsystems and materials considered highly likely. After the first flight the J-20B designation was more widely used by Chinese sources to refer to the new design – although this was still not officially confirmed.[19] It is expected that once the J-20S twin seater moves beyond prototype stages, the series-produced two-seat model will be based on the more up to date J-20B rather than the older J-20A.

FUTURE VARIANTS

In March 2018 speaking to *China Daily* Yang Wei highlighted plans to: 'develop the J-20 into a large family and keep strengthening its information-processing and intelligent capacities.'[20] Prominent figures connected to the program made multiple statements to similar effect, prompting speculation as to what a 'large family' could look like and whether this was an indication that multiple variants

In August 2020 reports emerged that the Chengdu Aircraft Design Institute was developing an AEW&C variant of the J-20, and subsequently in January 2022 the chief editor of Beijing-based magazine *Aerospace Knowledge*, Wang Ya'nan, highlighted that a J-20 with a larger radar and new fire controls could be fielded as a small AEW&C system.[22] Such an aircraft could serve as a command post for other units with special communications equipment and data links. It remains uncertain how heavily modified the J-20S would be for AEW&C roles, as while its radar is already one of the largest on any fighter in the world, modifications to accommodate an even larger sensor suite remain possible. Heavyweight fighter/interceptor airframes have been used in AEW&C roles in the past, capitalising on their high endurances and large sensors, with the Soviet MiG-31 and Su-27PU interceptors both developed with the ability to serve as command posts. A more radical unrealised Soviet program was the Su-27K AEW&C – which would have seen a 360° search radar with antenna integrated above the fuselage of a Flanker in a revolving 'saucer' radome.[23]

The J-20's potential to serve as a strike fighter was widely highlighted in Western assessments beginning as early as 2011, with a 2015 RAND Corporation report observing that the fighter's 'combination of forward stealth and long range could hold U.S. Navy surface assets at risk, and that a long-range maritime strike capability may be a cause for greater concern than a short-range air-superiority fighter like the F-22.'[24] A U.S. Naval War College report similarly highlighted the value of an 'effective surface-attack platform for out to several hundred nautical miles at sea' – originally in the belief that the J-20 was not primarily designed for air superiority.[25] While J-20s are expected to be able to launch a range of cruise missiles, the possibility of developing a specialised derivative which sacrifices air-to-air performance for an expanded payload and increased range remains. With the U.S. Defense Intelligence Agency's annual report on the PLA concluding in January 2019 that China was 'developing new medium- and long-range stealth bombers,' a modified J-20 could serve as a medium range bomber complementing the intercontinental range H-20.[26]

Developing a heavily modified strike fighter based on the J-20 would hardly be unprecedented, with the F-22's primary contractor Lockheed Martin having from 2002 studied the possibility of developing a derivative with double the range and a significantly increased internal payload – a design later referred to as the FB-22.

The aircraft had a reduced acceleration, speed, and manoeuvrability, but could carry 375 percent the ordinance

JH-7 strike fighter flying through a valley during at very low altitude during a training exercise. (China Military Online/Yang Pan)

and 80 percent more fuel while its fuselage remained relatively unchanged.[27] Persistent issues with the F-22 were used to argue against the program, however, and it never moved beyond conceptual stages.[28] The Su-27 was also developed into a strike variant, the Su-34, which was approximately 50 percent heavier and had a tremendously increased range and weapons payload at the expense of air-to-air performance.[29] A J-20 strike variant, possibly designated JH-20, could follow much the same path, and would increase commonality between Chinese tactical combat air units while providing considerable economies of scale to the J-20 program and potentially replacing approximately 260 JH-7 strike fighters. With J-20s already having significantly longer ranges than that the proposed FB-22, a similar extended range variant would allow for very prolonged operations across much of the Pacific, the Indian Ocean and the Arctic. While developing such an aircraft would align with the PLA's current known priorities for modernisation, there have been no official indicators that such a variant has been considered.

The development of electronic warfare or drone-controller aircraft based on the J-20S has been widely speculated since before the twin-seater was itself confirmed. In April 2021, amid persistent rumours that a twin-seat variant would imminently be unveiled, defence experts writing for the Xi'an-based military magazine *Ordnance Industry Science Technology* highlighted regarding the advantages and roles it could have: 'The emergence of a twin-seat version of the J-20 is because the J-20's mission has diversified and China needs a more capable fighter jet... It's a piece of cake for the J-20 to perform [electronic interference] duties because of its strong power supply capacity, fire-control radar and integrated avionics system.' For electronic attacks: 'We can imagine that the front pilot will be in charge of flying the aircraft, while the pilot sitting behind is in charge of controlling the electronic inference platform, making the J-20 a nightmare for enemy electronic equipment.' The second seat could accommodate a controller for drones, as the experts noted: 'The drones could be bait to attract enemy aircraft or draw in stealth aircraft ... they can also gather intelligence, carry out attacks against air defence systems and gain air superiority.'[30]

Wang Ya'nan, chief editor of Beijing-based magazine *Aerospace Knowledge*, highlighted in an interview in January 2022 that a rear seat officer in a twin-seat fighter could help facilitate more complicated combat operations, including by serving as a drone controller. Drones could accompany J-20 units in-flight and expand a unit's firepower, situational awareness and electronic warfare capabilities considerably.[31] In October 2022 state media outlet CCTV unveiled computer graphics showing J-20s commanding formations of flying stealth drones, possibly based on the GJ-11 airframe, which was expected to be a defining feature of sixth generation fighters and which the J-20, as the only fifth generation fighter with a twin-seat variant, was well optimised to doing.[32]

FUTURE TECHNOLOGIES

Much like the F-35, the Flanker, and other fighters developed under larger programs, the J-20 is expected to continue to see substantial improvements to its performance made over its years in production and beyond as China's defence sector and its tech sector more broadly strengthen their leadership in developing key emerging technologies. Since shortly after the aircraft first entered service, J-20s across the fleet have received continuous software upgrades as successive batches have been produced with increasingly advanced capabilities, while stealth capabilities and a range of subsystems have made significant incremental improvements between batches resulting in major performance advantages for newer models. Chinese officials have highlighted since the first J-20s became operational that control systems, stealth coatings, hull materials, sensors and engines were key areas where work was being done to improve the aircraft.[33]

J-20 at the Zhuhai Airshow in 2018 with weapons bays open showing PL-15 and PL-10 missiles. (Colin Cooke/ Openverse)

to highlight the possibility of laser weapons integration.[39]

Facilitating greater network-centric operations, National People's Congress deputy and J-20 pilot Gao Zhongqiang in April 2023 highlighted that the fighter's communications could be a key focus for future enhancements in order to facilitate more seamless network-centric operations. 'A pilot should jump out of his own perspective in aerial combat... and put himself in a system of joint operation, contributing more to winning as a whole,' he stated, stressing that while operating the J-20 and other modern fighter classes he had already become accustomed to receiving a vast cloud map of data. He called for greater integration between the different services of the PLA through a constantly improving communications network – with the publication of his statement in state media indicating it had a degree of endorsement.[40] The network Gao called for resembled projections in the U.S. of what a sixth generation battle network with fully seamless information sharing between platforms across the services would look like, fitting in with the Joint All-Domain Command and Control (JADC2) concept.[41] As summarised by the *Global Times:* 'vast data will likely be pre-processed, selected and highlighted by artificial intelligence before being presented to human operators, so the latter can make the best decisions for victory.'[42]

In April 2021 the PLAAF pilot who had flown the J-20 on its maiden flight, Li Gang, raised the possibility of the fighter integrating thrust-vector controls across its production run for improved manoeuvrability.[43] There had long been speculation that the fighter had already integrated such engines, with chief designer Yang Wei stating in 2018 when asked 'You asked about when, but how do you know it hasn't?' Footage released in late August 2022 showed a J-20 demonstrating very high levels of low-speed manoeuvrability at the Changchun PLAAF Open Day in Northeast China, with its extreme sharp climbs and turns contrasting with the basic flybys and conservative turns that previously demonstrated.[44] Footage released a month later showed unprecedentedly complex and strenuous low-speed manoeuvres, including turns during vertical climb and gliding while nose up, fuelling further speculation that thrust-vectoring engines had been integrated.[45] Further footage of extreme manoeuvres was released in November, with more subsequently continuing to appear.[46]

The J-20B with WS-15 engines was expected to integrate thrust-vectoring engines across its production run. With a fast-growing portion of the fleets of China's potential adversaries made up of stealth aircraft, the mutual difficulties opposing stealth jets would have targeting each other at beyond visual ranges made short range engagements more likely, with the extreme low-speed manoeuvrability provided by thrust vectoring being particularly advantageous at such ranges. Russian sources further claimed thrust vectoring allowed fighters to 'disappear from the locators of enemy

Beyond these incremental improvements, advances in a number of new technology areas have the potential to more radically change how the J-20 can contribute to military operations. Although opacity surrounding military R&D efforts has left many of Chinese aviation's potential new capabilities to be fielded to speculation, available information provides some important indications. A widely expected early improvement is the increasing of the fighter's internal air-to-air missile payload, with this possibility highlighted in multiple Pentagon reports and being prioritised for America's own stealth fighter programs.[34] Since the J-20's central weapons bay was first seen observers highlighted that it was 'remarkably uncluttered' and could potentially house more missiles.[35] One means to achieve this would be integration of new missile racks, as seen on the F-35A/C with the addition of 'Sidekick' racks.[36] A trapeze launcher system similar to that of the F-22 could also be key to achieving this – although the possibility remains that such a solution may have already been implemented since the contents of J-20 weapons bays were rarely revealed. Alongside increases in the quantity of missiles carried, superior new variants of the PL-10 and PL-15, and a next generation of more capable missiles, were all anticipated. According to Pentagon reports this included a ramjet-powered long-range missile, providing a larger 'no-escape zone' and thus an improved long range kill probability particularly against manoeuvrable targets.[37] The PL-10's successor was sometimes informally referred to as 'PL-16,' although there were no indications as to what capabilities it could field.

Beyond improvements to missile capabilities, state broadcaster CCTV reported in 2020 that an airborne laser attack pod was under development. Chinese defence firms had publicly displayed directional-emission high-energy laser anti-aircraft weapons like the LW-30 at exhibitions.[38] Lasers left no time lag after firing for targets to evade or activate countermeasures, and potentially allowed fighters to engage more targets per sortie than they could with a limited missile arsenal and at a much lower cost per shot fired. The fact that they required a large amount of energy to be effective, however, was expected to significantly delay their integration onto fighter-sized aircraft. The *Global Times* was among multiple sources

J-20A makes first public demonstration of high manoeuvrability at the Zhuhai Airshow in 2018. (China Military Online/Wang Weidong)

fighters' by remaining still momentarily and thus discarding radar capture.[47]

A number of stealth technologies under development in China showed significant promise to improve the J-20's survivability. Among these was the work of a team at the Chengdu Institute of Optics and Electronics to create the world's first mathematical model to precisely describe the behaviour of electromagnetic waves when they strike a piece of metal engraved with microscopic patterns. Metamaterials were fine surfaces engineered on a nanoscale that were theoretically capable of bending infrared radiation, with a breakthrough in 2019 facilitating the creation of 'meta surface' membranes which could absorb radar waves in the widest spectrum yet reported at between 0.3 and 40 gigahertz.[48] Reports from August 2022 highlighting that a plasma device had been developed to improve combat aircraft's radar absorption, after a similar system using whips of plasma to reduce RCS had been developed in the USSR for the MiG 1.42 but never completed.[49]

Breakthroughs made in China also held promise of revolutionising the J-20's ability to engage other stealth aircraft, with the *Infrared And Laser Engineering* journal publishing in 2022 that an IRST developed by an unconfirmed division of China's defence sector had provided clear images at a 285km range – with the outline of the target aircraft including its rotor, the tail and number of engines all clearly visible.[50] Due to the limited number of photons which can reach a detector's lens over extreme distances IRSTs have generally had ranges of under 120km, which restricts them to operating as secondary sensors, with such sensors providing a valuable means of tracking low RCS targets using their heat signatures.

The USAF notably sought to allow the F-35 to contribute to missile defence operations by 2025,[51] after the fighter first successfully tracked a missile launch and transmitted tracking data in 2014 and improved significantly on this capability over subsequent tests.[52] After the U.S. withdrawal from the Intermediate Range Nuclear Forces Treaty in 2019, which was expected to pave the way for major surface-to-surface missile deployments across the Western Pacific, the value of such a capability for the J-20 increased significantly. Like the F-35, being deployed widely and integrating powerful sensors allowed the J-20 to potentially significantly increase the situational awareness of missile defences and provide cueing data to interceptors, with both fighters speculated to potentially also integrate anti-ballistic or compact solid state airborne laser weapons to themselves target missiles in their early stages.[53]

THE NETWORK-CENTRIC AGE

The 2000s and 2010s saw all aspects of aerial warfare become increasingly network-centric, with the rate at which this has occurred perhaps best demonstrated by a comparison of the F-35 and J-20's avionics with those of the older F-22. The newer fighters are capable of sharing data and fusing sensor inputs with all manner of assets from satellites to accompanying combat aircraft, while the F-22 struggles to share data even with other fighters requiring accompanying 'translator' drones which began to be fielded only in the early 2020s.[54] As observed in 2016 by former PACAF Vice Commander Lieutenant General (ret.) David Deptula regarding how fifth generation fighters differed from their predecessors, they 'are not fighters... they are flying sensor-shooters that have the ability to act as information nodes in a combat cloud – a universe if you will – made up of platforms, not just airborne but also operating at sea and on land, that can be networked together.'[55]

The increasingly network-centric way aerial warfare assets operate means the capabilities of top-end fighter units has increasingly been determined not only by the characteristics of the aircraft themselves, but also by those of friendly assets in their surrounding networks. The potency of the J-20 in the PLAAF can thus only be fully comprehended with an understanding of some of the range of complementary assets which will operate alongside it. As summarised by Air Force Secretary Frank Kendall in 2022: 'it's not just about fighter [versus] fighter. It's a much more complicated equation than that.'[56]

In August 2016 shortly after the first J-20s were delivered to the PLAAF, when asked how the new fighter would compare to the F-35, USAF Chief of Staff General David L. Goldfein stressed that this was far less relevant in an when fighters networked with a range of supporting assets under broad national defence networks. His answer provided one of the best summaries of how network-centric operations had transformed aerial warfare. Raising the example the F-117s he previously flew, which joined the USAF in 1983, Goldfein stated:

> It was single domain, it was a closed system and it was a sequential way of applying air power because I was always going to be out in front of anybody else on the ground or at sea. The F-35, now since you're asking about F-35, J-20 is a completely different mindset. It starts talking in the network before the pilot even climbs the ladder. It starts comparing information, it starts placing symbology on the visor of the pilot.
>
> That symbology is replicated not only in the displays but across the network of everywhere it's joined. So when we apply fifth generation technology, it's no longer about a platform, it's about a family of systems and it's about a network and that's what gives us an asymmetric advantage so that's why when I hear about an F-35 versus J-20, it's almost an irrelevant comparison because you've really got to think about a network versus a network. This is combat in the information age.[57]

While during the Korean War fighters fought at visual ranges and carried no radars or missiles, relying primarily on ground-based radars and control centres for support, by the twenty-first century it was increasingly vital for fighters to be able to network with assets such as satellites and AEW&C systems – with these providing targeting data, routing communications, and guiding weapons among other functions. Thus unlike its direct predecessor in the PLAAF's elite units the MiG-15, the J-20 for all its strengths could not come close to guaranteeing Chinese security in the air domain without support. A range of complementary assets were thus developed for the PLAAF in parallel to the J-20 program, with their performances being as critical to determining the fighter's operational effectiveness as its own intrinsic capabilities were.

Between the J-20's unveiling in 2010 and its service entry in 2017 the standing of China's defence sector, and the capabilities of top PLA frontline units particularly in the Navy and Air Force, improved tremendously. The world's first quantum communications satellite,[58] the world's most capable missile destroyer,[59] the world's second largest fleet of cutting-edge AEW&Cs,[60] the world's largest military transport aircraft in production,[61] and a new generation of missiles and sensors for fighter jets, were among the examples of programs which marked the PLA and Chinese defence sector's ascension from unremarkable medium level players to a close contender for global leadership. As late as 2010 many of these capabilities were either entirely unanticipated, or were expected to be realised considerably slower, on a smaller scale and with less cutting-edge features. Thus, while China's armed forces in 2010 would have been poorly suited to properly introduce and effectively operate a fighter as sophisticated as the J-20, its transformation over the decade had no parallels among other major militaries and ensured it could operate effectively as part of broader networks of similarly advanced assets.

MANNED SUPPORTING AIRCRAFT

Although twenty-first century fighters are capable of operating far more autonomously from radar and control units than their Korean War-era predecessors, AEW&C systems nevertheless continue to be considered key force multipliers. It was thus highly notable that when the commander of the U.S. Pacific Air Forces General Kenneth Wilsbach confirmed in March 2022 the J-20's first encounter with F-35s, he particularly singled out the strong impression made by the Chinese fighter's associated command and control and in particular the supporting role played by the PLAAF's top AEW&C system the KJ-500. 'We're relatively impressed with the command and control associated with the J-20,' Wilsbach stated, observing that 'aircraft in the region... were pretty well commanded and controlled by the Chinese assets.'[62] 'The KJ-500,' he added, 'plays a significant role in some of their capability for long range fires. Some of their very-long-range air to air missiles are aided by that KJ-500. Being able to interrupt that kill chain is something that interests me greatly.'[63] Although under four years into high-rate production, approximately 25–28 KJ-500s were thought to be in service at the time, making it by far the most widely fielded AEW&C platform outside the U.S. Armed Forces. Experts cited by the *Global Times* seven months

KJ-500 AEW&C system. (China Military Online/Gao Hongwei and Zhang Bin)

KJ-500 AEW&C systems. (China Military Online/Qin Qianjiang and Zeng Qi)

AKJ-500 AEW&C system. (航空君 on Weibo)

later highlighted that pairing the KJ-500 with the J-20 would help to ensure an advantage even over other fifth generation units when operating near Chinese territory.[64]

After observing the key advantages USAF E-3s provided the service's fighter units in Operation Desert Storm, where they assisted in 38 of the USAF's recorded 41 air-to-air kills,[65] the PLAAF which had no AEW&Cs itself in the 1990s prioritised development of a similar capability. This first materialised in the KJ-2000 system commissioned from 2004.[66] Fifteen years later China's AEW&C fleet had grown to over 40 aircraft, with a major landmark being the induction of the KJ-500 into service from 2015. The aircraft introduced new levels of sophistication and brought Chinese AEW&C capabilities up to the cutting edge. Production was expanded in 2018 to well over double the rates of preceding Chinese AEW&Cs,[67] before the enhanced KJ-500A variant was unveiled in 2022. The KJ-500's induction appears to have strongly influenced the USAF's decision on how early to retire its very large but increasingly obsolete E-3 Sentry fleet and replace them with modern E-7 Wedgetails – a technological equivalent to the new Chinese aircraft. The decision was finalised in 2023, after officials including General Wilsbach repeatedly stressed both the extreme limitations of the E-3 and the high sophistication of new Chinese AEW&Cs.[68]

The PLA places a much lower emphasis on power projection than leading Western militaries, and its fighter units have very rarely operated far beyond China's borders, which has made them significantly less reliant on aerial refuelling. The PLA's ratio of heavyweight long-range fighters to lightweights is also far higher than that of any Western air force – over double that of the USAF – meaning the average Chinese fighter has a much higher endurance than the average U.S. or NATO fighter. Furthermore, even among heavyweights, the J-20 benefits from a very comfortably greater range than its Western counterparts with over double the range

YY-20 refuels J-20s and J-16s during flight at the 2023 Air Force aviation opening event and Changchun Airshow, 26 July 2023. (China Military Online/Yan Linyun)

J-16D electronic warfare aircraft. (耿直的鲁斯兰 on Weibo)

of the F-15 and F-22. Thus, while both aerial tankers and external fuel tanks are considered indispensable for Western air operations particularly in the Pacific, Chinese fighters can operate effectively without either.

The PLA at the beginning of the 2020s fielded an estimated 16 tanker aircraft, 12 of them very small HY-6s based on the H-6 bomber airframe, which was a very modest fleet even by the standards of much smaller countries. The heavily power-projection oriented U.S. Military by contrast fielded over 450, which were on average much larger and longer-ranged.[69] Despite the J-20's already very high endurance, when the very-large long-ranged YY-20 joined the PLAAF in late 2021 it had the potential to transform its scope of operations by enabling extended patrols as far as Guam, the Arctic and the Indian Ocean.[70]

Although advanced AESA radars provide significant electronic warfare (EW) capabilities, as do electronic warfare suites and pods, specialist EW aircraft can provide more potent support and serve as force multipliers for fighter units. The J-16D EW aircraft was accordingly developed with antennas and EW arrays built into its fuselage and integrated jamming pods able to analyse radar frequencies and locate radar transmitting devices. Inheriting the original J-16 fighter's high flight performance and high endurance, it could operate alongside J-20 units even on longer-ranged missions. The only equivalent aircraft fielded abroad was the EA-18G Growler serving in the U.S. Navy and the Royal Australian Air Force. As the J-16D entered service late in 2021 the *Global Times* emphasised that its purpose was largely 'to team up with the J-20,' as fielding them together 'would bring tremendous combat efficiency.'[71] Eventual development of a specialist EW aircraft based on the J-20S airframe was also considered possible. Further EW support is provided by other aircraft such as H-6G bombers, which have a more defensive focus and are fielded primarily to protect friendly assets across wide areas from enemy electronic attacks using their countermeasures suites. Their mission extends to disrupting enemy communications and connections with satellite assets, with militaries across the world particularly valuing such disruptive capabilities from 2022 after their importance was demonstrated by Russian EW operations in Ukraine.[72]

UNMANNED SUPPORT AIRCRAFT

The PLA from the early 2000s strongly prioritised modernisation of its UAV capabilities,[73] with the industry having by the mid-2010s increasingly shown signs of leadership placing China in a league of its own with the United States comfortably ahead of other competitors. The PLA by the end of the decade fielded what was by many estimates the widest array of high-performance medium and heavyweight drone classes of any military worldwide, many of which had no foreign equivalents. Notable examples included the

WZ-8 surveillance platform unveiled in 2019, which was the world's first operational hypersonic aircraft, as well as the large subsonic Divine Eagle which deployed a wide range of sensors from high altitudes. New generations of elevated sensor and reconnaissance UAVs such as the high altitude WZ-7 have continued to complement the PLA's growing satellite and AEW&C capabilities, as well as fighters' increasingly advanced radars and data links, and as a result situational awareness of PLA battle networks is expected to continue to improve rapidly with significant implications for how the J-20 can operate.

J-20S twin-seater prototype. (飞扬军事铁背心 on Weibo)

An emerging trend in the use of unmanned aircraft is the pairing of manned fighters with unmanned 'wingmen,' with future iterations of the J-20, F-35 and Su-57 all expected to be integrated with such support as force multipliers. Between four and six UAVs accompanying each fighter will be able to use their sensors to maximise situational awareness, their armaments to allow each pilot to direct several times more firepower, and their likely superior stealth capabilities and relative expendability to deploy closer to enemy defences. They are also expected to carry supporting equipment such as electronic warfare suites and directed-energy weapons, and some are expected to serve specifically as decoys to draw fire and thus protect their higher value 'teammates.' The result will be fleets which are overall significantly larger but have smaller numbers of manned fighters, as approximately 80 percent of new generations of units are comprised of unmanned 'wingman' aircraft.

GJ-11 flying-wing stealth drone. (秦浩博 on Weibo)

'Wingman' UAVs are expected to begin entering service in the mid-late 2020s, and are expected to initially be semi-autonomous, but gradually gain near full autonomy. With much greater funding for AI R&D, a range of metrics have strongly indicated that China has gained a lead in AI development ranging from the quantities of papers published to the state's unique ability to apply the technology for innovative responses to national emergencies.[74] This has positioned the defence sector well to make the J-20 a world leader in operationalising manned-unmanned pairing. Development of the J-20S twin-seater, which uniquely for its generation can accommodate a drone control officer, further strengthens the program's position.

With the potential to support J-20s in air-to-air combat at all ranges, the Dark Sword program was expected to operationalise a fighter-sized UAV in the late 2020s. Using a diverterless supersonic inlet, the canard twin tailed aircraft appeared designed to emphasise high manoeuvrability and effective employment of anti-aircraft missiles for high-end air superiority missions. The possibility of the Dark Sword eventually evolving into an unmanned autonomous next-generation fighter in its own right was widely raised when the design was first unveiled. A more conventional fighter-like airframe that also appeared optimised for air-to-air combat and similarly emphasised stealth and manoeuvrability was developed under the FH-97A program, with the UAV shown in computer-generated presentations operating networked with J-20s and flying in formation both to penetrate air defences and to engage enemy fifth generation aircraft.

A range of other unmanned aircraft were candidates to serve as 'wingmen' for the J-20, including the stealthy GJ-11 flying-wing UAV which could potentially carry sensors and missiles for beyond-visual-range air-to-air combat, and the more mysterious CS-5000T unveiled in November 2022 which also used a flying-wing stealth design and had an entirely unknown purpose. The WZ-10 Cloud Shadow appeared intended to provide electronic warfare support with a secondary armament of missiles and was also speculated to potentially be intended as a 'wingman' for fighter units.

As the J-20's communications suite and data links continue to be modernised its ability to synergise and share data with unmanned aircraft will only grow, including not only 'wingmen' nearby but also supporting aircraft much further away such as Divine Eagle elevated sensor drones and WZ-7 and WZ-8 reconnaissance platforms. Advances in artificial intelligence will be key to improving coordination. Integration with UAVs is expected to be one of the key developments that increasingly sets the J-20 and F-35 apart from baseline fifth generation fighters like the F-22, bringing them to a level that may come to be termed the '5+ generation.'

'J-20 DNA' IN THE FOURTH GENERATION

Although the air forces of many medium-sized countries very often invest in acquiring just a single class of fighter at a time, with this applying to most European states whether France with the Rafale, Turkey with the F-16 or Sweden with the Gripen, such fleets come with greater limitations and have in almost all cases relied on lightweight fighters with relatively small radars and low endurances. In contrast, for countries seeking to benefit from the complementary performance advantages of different classes, including those seeking to field high-performance heavyweight fighters which would be unaffordable to operate in large numbers, acquiring multiple fighter classes in parallel has been common. Notable examples were the USAF F-15 and the Navy's F-14 acquired in limited numbers as part of high-low pairings with lighter F-16 and F-18 fighters. Only the U.S. and USSR initially produced complementary aircraft for such pairings, the latter with the Su-27 and MiG-

Heavyweight J-11B and lightweight J-10C fighters. (China Military Online)

Heavyweight Su-27 and medium-weight MiG-29 fighters. (Wikimedia Commons/Dmitry Terekhov+)

29, before being joined by China in the 2000s with the J-11 and J-10. A French attempt to field a similar pair failed in the 1980s as its heavyweight Mirage 4000, developed as a complement to the lightweight Mirage 2000, proved unaffordable.

While the USAF for most of the 2010s had no fighter other than the F-35A on order, this was a result of both issues with the F-22, and the fact that a sixth-generation heavyweight fighter was still relatively early in its development. The U.S. Navy, meanwhile, abandoned a high-low combination after the Cold War to field a fleet with much lower operational costs comprised exclusively of medium-weight F-18E/F fighters – albeit at serious detriment to the protection they could offer carrier groups.[75] In the Russian Air Force, too, almost all new fighters received in the first 30 years after the USSR's disintegration were Flankers, with the lack of any meaningful fighter acquisitions until the 2010s and contraction of the defence budget and fleet-size leading to abandonment of the high-low combination.

A contrasting trend was seen in China where the diversity of the fleet and of production lines grew significantly after the Cold War's end, with the J-20 accordingly developed, produced and commissioned by the PLAAF alongside two complementary fighter classes: the Shenyang J-16 and Chengdu J-10C. Both were produced on far greater scales than any Russian or American fighter other than the F-35, facilitating modernisation of frontline units far faster than reliance on the J-20 alone would have allowed, with their most cutting-edge features based on technologies developed for the J-20. Significant commonalities ranged from the J-20's distinctive green heads-up display, to controls, cockpit displays, helmet-mounted sights and weaponry. Composite material technologies, data links and sensors were among

J-10C with PL-15 and K/JDC01A laser designator pod. (China Military Online/Wang Guoyun)

People's Liberation Army Air Force Su-35 Flankers. (空军在线 on Weibo)

The PL-10, PL-12 and PL-15 represent current generation of PLAAF air-to-air missiles, while the USAF, and some allies, employ the AIM-9X and AIM-120C. (Artwork by Tom Cooper)

December 2023, the first publicly released image showing PL-XX missile in active service with J-16 fighters. (China Military Online)

the multitude of other areas where tremendous advances had been made to realise the goals of the J-20 program, with these bringing the fighter industry significantly forward and being used to significantly enhance the J-10 and Flanker designs. China and the United States stood out for their introduction of such '4+ generation' fighters with fifth generation level avionics, like the J-16 or F-16E/F, in parallel to their first fifth generation fighters, where by contrast Russia and Europe had to rely heavily on '4+ generation' fighters for decades before their industries reached a genuine fifth generation level.

The J-10C radically improved on the original J-10A and repeatedly proved to be highly capable particularly in air-to-air combat. In June 2020 reports emerged that the fighter had won overwhelmingly in simulated combat against PLAAF Su-35s recently acquired from Russia, which remained the Russian Air Force's top operational fighters. This was a significant achievement considering that the J-10C was approximately half the Russian jet's size and was far from the PLAAF's top fighter. Reduced radar visibility reportedly played an important role in helping the J-10C 'achieve success in the confrontation with the Su-35' which, by allowing the Chinese fighter to detect and fire first, 'gave an advantage in the simulation of a missile attack.' The J-10C benefitted from a much lower radar signature and significantly more advanced sensors and weaponry,[76] while its PL-10 and PL-15 missiles had tremendous advantages over their Russian counterparts in both visual and beyond-visual ranges. In December 2021 the J-10C repeatedly gained victories against both the Su-35 and the J-16, marking the third consecutive year it had bested both in simulated engagements since entering service in 2018.[77]

The J-16 entered service in the mid-2010s as by far the world's most advanced Flanker derivative, and despite not being marketed for export was produced at at least twice the rate of any foreign heavyweight fighter, including Russia's own Flanker models. It was fourth in the world in terms of annual production rates, but second in the world among heavyweights, outproduced only by the J-20. Fellow and expert on Chinese security at the Royal United Services Institute Justin Bronk referred to the fighters as 'genuinely better in almost every important regard than their Russian Flanker counterparts... not only do you have something that's lighter and stronger than a Russian Flanker, that has access to far more capable multirole munitions both in the air to ground domain... [but also] much more effective air-to-air weaponry to go with that AESA radar.'[78] For longer-ranged missions the J-16's high payload, range and altitude allowed it to keep up with the J-20, while its PL-XX air-to-air missiles, which were too heavy for the J-10C and too large for the J-20's missile bays, allowed it to seriously threaten support aircraft such as AEW&Cs and aerial tankers vital to Western air operations.[79]

Pentagon reports in 2020 highlighted that the J-20 could prove particularly challenging to tackle when operating alongside the J-16 and J-10C and with KJ-500 support,[80] with the three fighters demonstrating highly complementary capabilities. The two '4+ generation' fighter programs were key to ensuring that the J-20, which even with a high production rate still equipped only a small portion of PLA air brigades, would operate in a fleet where squadrons with similarly advanced weaponry, sensors and network-centric warfare capabilities were widely fielded. Two further fourth generation fighter classes developed for service outside the PLAAF benefitted from many of the same technologies, namely the J-15B built for the Navy's carrier operations and the very light JF-17 Block 3 developed for export, with the former's induction and latter's proliferation facilitating greater ease of joint operations with other services. Similarly advanced avionics and weaponry were also from the late 2010s increasingly widely integrated onto older J-10 and J-11 fighters as part of upgrade packages which provided a cost-effective means of more widely bringing key aspects of performance up to the latest standards.

5
INTERNATIONAL CHALLENGES AND RESPONSES TO THE J-20

THE J-20 AS A CHALLENGE TO AMERICAN POWER

By the early 2020s a strong rivalry had emerged between the J-20 and F-35 programs, which stood out within their generation as the only ones with ongoing large-scale production. Both led the world in their production scales, had unique avionics not seen on other aircraft, and were rapidly enhanced with a succession of incremental upgrades not seen in other programs. It was thus a major landmark in the development of both programs when their first encounter was confirmed on 14 March 2022, to have occurred over the East China Sea, marking the first ever encounter between stealth fighters from opposing sides. Commander of the U.S. PACAF Air Force General Kenneth Wilsbach stated at the time regarding PLAAF J-20 operations: 'What we're noticing is they are flying it pretty well. We recently had – I wouldn't call it an engagement – where we got relatively close to the J-20s along with our F-35s in the East China Sea, and we're relatively impressed with the command and control that was associated with the J-20s.'[1]

Wilsbach's statement provided a rare insight into prevailing thought in the USAF on the Chinese stealth fighter, which had persistently surprised Western analysts over the past 11 years with

J-20A with PL-10s ready to fire and an open central weapons bay, turning left at the Zhuhai Airshow in 2018. (China Military Online/Wang Weidong)

An F-35A of 302nd Squadron, Japanese Air Self-Defence Force. (Wikimedia Commons)

both the speed of its development and the sophistication of its demonstrated capabilities. The PACAF chief's assessment followed a Defense Department report to Congress referring to the J-20 as having 'high manoeuvrability, stealth characteristics, and an internal weapons bay, as well as advanced avionics and sensors providing enhanced situational awareness, advanced radar tracking and targeting capabilities, and integrated EW systems.'[2]

As the world's only fifth generation air superiority fighter both in production and fielded at squadron level strength, the J-20's position indicated the Chinese fighter industry had succeeded in seizing a leading position internationally. The lighter F-35 was much better optimised for air-to-ground missions, the F-22's production had been cut very early, and the Su-57 was not expected to see its first full regiment formed before mid-2024 – and with capabilities that were not at a truly peer level. Thus, as much as the J-20's strong international standing was a result of Chinese advances, it was equally a consequence of Russian and American shortcomings. The former had abandoned the MiG 1.42, with the much less ambitious Su-57 then delayed tremendously, and the latter had failed to develop the F-22 into a viable successor to replace the F-15 over a long production run or with anything near the capabilities initially envisaged. Thus while in the 1980s when the Soviet and American advantages over the rest of the world appeared insurmountable, serious impeding of fighter development from the 1990s, in large part due to sharp contractions to their defence sectors and wider industrial bases, provided China with a window to develop the J-20 into what was very possibly the world's premier fighter for air-to-air combat. The J-20's entry into service thus marked the first time since the Second World War that a fighter from outside the United States or Russia could be considered a serious contender for the title of the world's most formidable in an air superiority role.

While the J-20 and F-35 were in a league of their own in terms of both production scale and sophistication, America and its allies' ability to rely on the F-35 in the face of peer-level Chinese competition was limited. A leading cause of this was the tremendous performance issues the fighter faced, ranging from software issues which on occasions caused crashes, to an underpowered engine that limited potential to integrate upgrades, caused tens of billions of dollars in additional operational costs, and deeply slashed availability rates.[3] Despite being the newest airframes in the fleet, the F-35's high maintenance needs meant availability rates were rivalled only by the F-22 for their deficiency, with even F-15C/Ds built during the Cold War and operated decades longer than intended performing much better in this regard.[4] Progress in making the aircraft operationally viable was slow, with the number of flaws consistently remaining at around 800–900 for years while Lockheed Martin and the Pentagon failed to bring this figure meaningfully down.[5]

The poor state of the F-35 program was attested to by a number of sources from military think tanks such as NSN[6] and RAND,[7] to organisations such as the Project on Government Oversight,[8] and individuals such as the Pentagon's chief weapons tester Michael Gilmore[9] and Marine Captain Dan Grazier.[10] More bluntly, Defense Secretary Christopher C. Miller stated in 2020 that 'the F-35 is a piece of ****,' while his predecessor Patrick Shanahan referred to the troubled fighter in 2019 as 'f—ed up,'[11] with similarly harsh albeit less abrasively worded assessments widely made by leading officials in both the civilian and military leadership.[12] An inquiry by the Pentagon's Office of the Inspector General found that Shanahan's criticisms of the F-35 'were consistent with other comments about problems in the F-35 program made by other senior DoD officials.'[13] Acquisitions were also scaled down considerably, and while the Air Force was initially expected to acquire 110 F-35s per year this was reduced to 80, then 60, and finally just 48 aircraft.[14]

A further factor limiting the Pentagon's ability to rely on the F-35 in the face of a peer-level fifth-generation challenge was that it was far from well optimised to serve as an air superiority fighter. As indicated by the name of the program – Joint Strike Fighter – it was designed with advanced electronic attack, network-centric warfare and strike capabilities but was much more limited in the air-to-air domain. Like the F-16 it was built to succeed, the F-35 used only a single rather than twin engines, was intended to have much lower maintenance needs and operational costs than its heavier twin-engine counterpart, and was restricted to much lower speeds, altitudes and levels of manoeuvrability and to carrying a much smaller weapons payload and a smaller radar. Its flight performance limitations significantly reduced its maximum engagement ranges compared to heavier aircraft like the F-22 and J-20, which could fire missiles from much higher altitudes imparting far more energy upon launch.[15] As observed by former USAF Chief of Staff General Mark Welsh, the F-35 ...

U.S. Air Force F-22 (right) and F-35 fifth generation fighters at Fort Lauderdale. (1st Lieutenant Sam Eckholm/United States Air Force)

> ... was never designed to be the next dog fighting machine. It was designed to be the multipurpose, data-integration platform that could do all kinds of things in the air-to-ground arena including dismantle enemy, integrated, air defences. It had an air-to-air capability, but it was not intended to be an air-superiority fighter. That was the F-22.[16]

J-20 in yellow primer from first serial production batch, seen in late 2016. (机外停车Rabbit on Weibo)

USAF Combat Command Chief General Mike Hostage similarly warned: 'If I do not keep that F-22 fleet viable, the F-35 fleet frankly will be irrelevant. The F-35 is not built as an air superiority platform. It needs the F-22.'[17]

Hostage predicted that while the F-35 was not optimally suited to an air superiority role, the F-15 and its lighter counterpart the F-16 would be obsolete by 2024. With F-22 numbers very low and shrinking, availability rates remaining by far the worst in the Air Force, and investment in their modernisation in the 2010s remaining highly modest, the Raptor fleet's viability was increasingly left in question. The result was a heavy burden on the F-35 fleet, with fast-growing deployments of the aircraft to East Asia repeatedly alluded to as being a response to growing Chinese aerial warfare capabilities and to the expanding capabilities of the J-20 fleet specifically.[18]

Compounding the challenge posed by the J-20 was a significant shortfall in the size of the American fifth generation fleet compared to even Obama administration-era projections. Defense Secretary Robert Gates had in 2009 stated: 'By 2020, the U.S. is projected to have nearly 2,500 manned combat aircraft of all kinds. Of those, nearly 1,100 will be the most advanced fifth generation F-35s and F-22s. China, by contrast, is projected to have no fifth generation aircraft by 2020.' He added: 'By 2025, the gap only widens. The U.S. will have approximately 1,700 of the most advanced fifth generation fighters versus a handful of comparable aircraft for the Chinese.'[19] Not only did the PLAAF receive its first J-20s four years before 2020, but the American fifth generation fleet remained well below the strength projected.

Even by January 2022, just before J-20 deliveries were accelerated, the U.S. military had approximately 660 fifth generation fighters in service, all with very low operational readiness rates and over two-thirds of which were F-35s still years away from being ready for high intensity operations and suffering major performance issues.

Second FC-31 technology demonstrator. (9谢艺航6 on Weibo)

The J-20 fleet at the time was estimated at close to 200 fighters. Approximately 40 percent of the F-35s in service, or 189 aircraft, were highly troubled early production aircraft, issues with which were so widespread that they were not expected to ever be made viable for actual combat. Many of these were expected to fly for a small fraction of their intended service lives due to structural defects, which made the discrepancy in numbers significantly narrower still.[20]

The numerical balance was set to be far less favourable for the U.S. by 2025, with significant expansion of the scale of J-20 production set to allow the PLAAF to receive the aircraft at a close to twice the rate at which the U.S. Air Force, Navy and Marine Corps combined received F-35s, with deep cuts to F-35 deliveries for 2023, and significantly deeper cuts in 2024, only exacerbating this. The expected deliveries of FC-31 medium-weight fifth-generation fighters to the PLA Navy by then would make China the first country to ever serially produce two post-fourth generation fighter classes simultaneously, further expanding its stealth fleet's numerical advantage in a potential East Asian war. The first indications that the FC-31 program may have put a fighter into production came in April 2022, a month after confirmation of plans to imminently begin retiring F-22s from service.[21]

Although Western concerns regarding the J-20's capabilities grew significantly as the program matured, concerns began to be raised as early as 2010 regarding what the program's success could signify for continued American air supremacy. Following the publication of the first images of the first technology demonstrator, China military scholar Richard Fisher highlighted that the J-20 would likely be able to outperform the F-22 particularly in thrust and non-afterburner speed, stressing that the U.S. 'should be reviving production of the F-22 and not just reviving production, we should be developing an advanced version of the F-22' to respond to the challenge. Noting the significant issues affecting the F-35 program, Fisher further warned: 'the F-35 needs another rework. It needs to be made competitive with this fighter [J-20].' 'Since World War II, the American military has never gone into battle without the assurance of air superiority. China is a rising power, and it is determined to challenge the American position globally. This fighter will allow them to do that on a military level… and from my perspective, that's simply unacceptable,' he concluded. Such assessments were already widespread at the time.[22]

Before the J-20 entered service Pentagon reports highlighted that the PLAAF could use stealth technology as a core component to transform 'a predominantly territorial air force to one capable of conducting both offensive and defensive operations.'[23] In 2014 the U.S.-China Economic and Security Review Commission described the J-20 in its report to Congress as 'more advanced than any other fighter currently deployed by Asia Pacific countries.'[24] A U.S. Naval War College report highlighted that year that upon entering service, the J-20 would 'immediately become the most advanced aircraft deployed by any East Asian Power.'[25] These assessments mirrored those made in China itself, with an early example being PLAAF Senior Colonel Shen Jinke's statement just weeks after the first J-20s were delivered that the new fighter class was 'developed to meet the needs of the future battleground' and would 'further enhance the overall combat capability of the Air Force.'[26]

TRAINING TO FIGHT THE J-20

China's challenge to American air superiority epitomised by the J-20 was responded to not only with decisions regarding research and development and other acquisitions related investments, but also by how the U.S. military trained for future air wars. The beginnings of this could be observed shortly after the fighter began to enter service. In the first week of December 2018 a full-scale J-20 mock-up funded by the U.S. Marine Corps was seen at the USAF Air Dominance Center in Savannah, Georgia, and according to the *Marine Corps Times* had been built for visual and sensor training.[27] The attempted replica differed markedly from the original, with control surfaces, exhaust nozzles and landing gear being the most noticeable inaccuracies.[28] A second more accurate mock-up was first seen in mid-2023 and was confirmed to have been used in Marine exercises in North Carolina in October that year.[29] Perhaps more importantly than the mock-ups' military significance was their indication, possibly intentionally as a signal to China, that the J-20

F-35A in J-20-mimicking colours at the Activation Ceremony of the 65th Aggressor Squadron with Lieutenant Colonel Brandon Nauta. (Airman 1st Class Josey Blades/U.S. Air Force)

65th Aggressor Squadron F-35 in J-20-mimicking colours. (U.S. Air Force/ Sgt Lawrence Sina)

was increasingly being factored into American training and war plans as the 'threat of tomorrow' as the program matured.

In May 2019 the USAF announced plans to reactivate its 65th Aggressor Squadron – a unit intended to simulate the capabilities of enemy fighters to provide dissimilar 'red on blue' air combat training. The announcement followed scathing reports, including a post-audit report in April from the Department of Defense Inspector General, on the state of pilot training and in particular its lack of relevance to facing twenty-first century challenges.[30] The aggressor unit was activated in June 2022 and based at Nellis Air Force Base in Nevada, which was the USAF's leading facility for test training. The decision to equip the unit with F-35As provided a clear indication from the outset that provision of training against stealth fighters would be a primary role for the 65th, and with the J-20 being the world's only non-American stealth fighter active on any meaningful scale there was little doubt that it was the primary target. The fighters flown by the 65th had long provided an indicator of which fighter classes the USAF saw as the most pressing challenge, with the unit having flown F-15s from 2005 to simulate the capabilities of the Su-27 and its derivatives, and previously the F-5Es to represent Soviet MiG-21s. When reactivated in June 2022, the 65th Aggressor Squadron's F-35s were shown in a colour scheme closely resembling one used on the J-20, mirroring the squadron's F-15s' prior use of a colour scheme derived from that of Russian Su-27 units. This again signified the J-20's position, and that of China, as the new perceived primary challenger to American air superiority: a position previously held by Russian and Soviet MiGs and Sukhois for over 60 years.

On 22 June 2020, the Commander of Air Combat Command General James M. Holmes suggested that F-22s could be used in aggressor training – with the F-35's suitability to simulate the J-20's capabilities having been questioned on the basis that it was a much lighter single-engine fighter designed primarily for strike missions.[31] Two months later in August further reports emerged of the U.S. military's preparations for the possibility of air-to-air

F-35As from 65th Aggressor Squadron in J-20-mimicking colours. (Tech Sergeant Alexandre Montes/U.S. Air Force)

65th Aggressor Squadron F-15C in colours mimicking the Su-27. The squadron flew F-15C/Ds until 2014. (Jamie Hunter)

combat with enemy stealth fighters, with the USAF beginning to use F-117 stealth jets to refine counter-stealth tactics.[32] The F-117 had been retired from service in 2006 and was flown by private contractors, which made the class more readily available for training than F-35s or F-22s.

The USAF's leading dissimilar combat training exercises – Red Flag – which are held at Nellis Air Force Base pitting 'blue' pilots against 'red' aircraft simulating their adversaries, from August 2021 placed an unprecedented focus on training for combat between fifth generation fighters. Commander of the 64th Aggressor Squadron at Nellis, Lieutenant Colonel Chris Finkenstadt, stated regarding the exercises: 'Based on our focus toward great-power competition, we need to make sure that those [blue air] guys are ready, and we do that by presenting the best possible atmosphere we can.' An Air Force press release indicated that 'red' units won 'a lot of red air victories' in early simulated clashes – which was expected to change as counter-stealth tactics and techniques improved.[33] F-22s were notably included on 'blue' team tasked with tackling 'red' F-35s, with Captain Patrick Bowlds who flew Raptors at Red Flag stating: 'Having them [F-35s] on red air adds a level of complexity' to an already 'complex scenario.' 'When you have a stealth platform on red air, it makes our job a lot more difficult in terms of knowing where they are, how we are going to protect allied forces or protect points on the ground or whatever the mission set is at that point in time,' he elaborated, adding that 'it is challenging, even flying the Raptor, to have good [situational awareness] on where the F-35s are.'[34] This provided important insight into how engagements between fifth generation fighter units over East Asia could look.

Beyond exercises, acquisitions were also made for training purposes specifically to support an improved capability against Chinese stealth fighters. In November 2020, for example, the USAF began to use a training program developed by firms Red 6 and EpiSci that projected seemingly real AI-flown J-20 fighters onto pilots' augmented reality helmet-mounted displays.[35] The Air Force from 2022 also sought to acquire a stealth target drone with the following requirements: 'The 5th generation representative target suite should be able to provide a remotely-controlled, destructible asset with threat representative RF [radio frequency] Emissions, EA [electronic attack] Emissions, Radar Cross Section (RCS) signature, Infrared (IR) signature, and internally carried expendables.'[36] The service had previously relied on modified F-16s flown as drones – designated QF-16s – to simulate the capabilities of enemy fourth generation fighters for live-fire targeting.

While the U.S. Air Force made significant investments to adapt its training to respond to China's fast expanding fifth generation capabilities, these efforts were undermined by the significant pressure on the fighter fleet that resulted primarily from issues with America's own fifth generation programs. In particular, the tremendous delays and cost overruns affecting the F-22 and F-35, and acquisition rates of both at just fractions of planned levels, forced the Air Force to keep fourth generation airframes in service for far longer than intended. As these aged their operational costs and maintenance needs rose steeply. By 2023, with F-22s and F-35s fielded at a fraction of numbers previously projected, the fleet's F-15s and F-16s had long outlived their planned service lives – averaging 38 and 32 years old – with their ages only set to continue to increase unless the older airframes were simply retired without replacement.[37]

Lieutenant General Joseph Guastella was among several figures in the Air Force leadership to highlight that the fleet's rising age was taking a serious toll on crews' flying hours[38] – while for those units which had transitioned to fifth generation fighters these aircraft's much lower availability rates and higher maintenance needs meant their situations were often little better.[39] 'We're on a collision course,' he stressed, warning that fighter force readiness could 'fall off a cliff.' At the time of the publication of a Mitchell Institute policy paper addressing the issue, which was co-authored by Guastella in June 2023, former PACAF Vice Commander Lieutenant General (ret.) David Deptula was among figures to highlight that Chinese pilots, flying fighters which were on average several decades newer, were getting more flying hours than their USAF counterparts. Additional training 'makes a difference,' he stressed, with this giving the PLAAF an important edge in better preparing itself for warfare at a fifth-generation level.[40]

THE J-20 AND AMERICA'S NGAD SIXTH GENERATION FIGHTER PROGRAM

The speed with which the J-20 was developed, brought into service, expanded in production, and incrementally modernised, revealed much regarding the state of China's defence sector, and in turn added considerable urgency to the American Next Generation Air Dominance (NGAD) sixth generation fighter program. While the USAF and naval aviation had for three decades increasingly emphasised air defence suppression rather than air superiority, based largely on the assumption that with the Soviet defence sector no longer in existence peer-level challengers in the air would only further diminish with time, the J-20 program was key to stimulating a partial reversal of this trend. Although as late as the mid-2010s the Air Force leadership had indicated that a sixth-generation fighter would be focused largely on electronic warfare and air defence suppression,[41] this changed as the J-20 program matured.

During the 1990s work on the NGAD's direct predecessor, the Advanced Tactical Fighter program which produced the F-22, was significantly undermined by the sharp deterioration in Russia's defence sector and armed forces. The MiG 1.42 program, which had been well positioned to produce fighters earlier than and outperforming the F-22, had been made unviable by Russian decline, and while the Flanker had initially added urgency to the F-22's development it was clear that post-Soviet Russia's defence sector would be unable to produce a similarly formidable successor in the next generation. Without any serious challengers the F-22 suffered cuts and delays that largely appeared to stem from complacency, with technologies removed to cut costs ranging from infrared tracking systems to secondary radars. The USAF had to restructure the program several times after 1991 as Congress continued to cut funding – what *Air Force Magazine* referred to as 'programmatic turmoil.'[42] Indeed, the Cold War's end was widely speculated to be a reason why Lockheed Martin's more conservative F-22 design was chosen over the proposed 'F-23' designed by Northrop Grumman, which was a more expensive, stealthier, faster and more ambitious option.

Where the F-22's development suffered for lack competition, the NGAD program is expected to be stimulated by Chinese advances in tactical combat aviation which have been demonstrated primarily by the J-20 program. This appears likely to mirror how the program to develop the F-15, the F-22's immediate predecessor, saw its capability requirements made more ambitious and its development accelerated in response to the demonstrated performance of the Soviet third generation MiG-25, which was considered the top fighter/interceptor of the early 1970s and which no prior USAF aircraft was capable of reliably engaging.[43] Experts on the F-15's development described the program as 'energized by the appearance of the prototype MiG-25 Foxbat… the MiG-25 was built for sustained Mach 3 performance, smashed a number of world performance records between 1965 and 1977… the MiG25's extraordinary performance was an unpleasant surprise to the West, as the MiG-15 had been in 1950. The effect of the MiG-25 revelation was to increase the emphasis on air superiority in the F-X specification.'[44] Thus while the F-22 program's results were in many ways underwhelming, the F-15 by contrast had truly revolutionary advantages over its predecessors and was considered a leading achievement of American Cold War military aviation – allowing it to become the fighter with the longest production run in history.

The J-20 and its successor well under development in China would similarly energise the NGAD program 50 years later, with Chinese industry's tremendous successes with the J-20 having significant implications for its ability to develop a sixth-generation fighter smoothly, quickly and with truly cutting-edge technologies. This raised the stakes considerably and meant unlike the F-22, which had faced no peer competition, there was far less room for the NGAD to similarly underperform. As analysts at *The Drive* were among many to note in 2020 regarding the need for a very different approach to developing NGAD compared to the F-22 due to peer-level challenges from Chinese fighter programs: 'the reality is that the sustainability of American air superiority over its peers has become so dire that there really isn't a choice. It's literally become a factor of change or die.'[45] They later observed in 2022 that any confidence the Pentagon had regarding its ability to tackle the J-20 appeared to have 'less to do with the jet's actual capabilities or how it might be employed,' and instead was based on its confidence in the capabilities which American sixth generation fighters were expected to bring to the field.[46] Similarly singling out the J-20 alone for mention, the monthly journal of the Air & Space Forces Association stressed that a sixth generation fighter was needed specifically to counter two threats – China's J-20 and new generations of ground-based air defences.[47] Discussing the J-20 specifically, Air Force Chief of Staff Charles Brown emphasised that the fighter could force the USAF to 'lose sleep' if the service failed to quickly field sixth generation capabilities, stressing that this was critical to 'maintain our advantage' in the face of Chinese advances.[48]

Image that may show 611 Institute's sixth generation technology demonstrator. (沉默的山羊 on Weibo)

Commander of the Air Combat Command General Mark Kelly in October 2021 went as far as to compare NGAD to the Manhattan Project in its urgency and importance, stressing: 'I would like to have more of a sense of urgency and a whole-of-nation effort towards it.' The extent to which the U.S. military relied on presumed air superiority to operate meant this was hardly an exaggeration. Due to the imminent nature of challenges to American air power, failing to do so would leave the USAF 'on the other side of coming in second in air superiority.'[49] Having warned nine months prior that China could bring a sixth generation fighter into service first,[50] Kelly reiterated this uncertainty at the time.[51] The complacency characteristic of the 1990s and 2000s had dissipated largely due to the J-20 program's successes not only in its performance, but in how quickly and cost effectively it had been developed, the rate at which the design had been enhanced post-service entry, and the rapid advances in Chinese industry and high-tech which had supported it and were set to make future fighters far better still.

AN/APG-81 radar antenna. (Wikimedia Commons/United States National Electronics Museum)

On 1 June 2022 it was announced that the NGAD program had entered its engineering, manufacturing, and development (EMD) phase, with Air Force Secretary Frank Kendall announcing that the fighter could be operational before 2030 and strongly emphasising the need to bring it into service quickly. Assessments at time highlighted that USAF plans to cut the number of planned operational squadrons, and to begin very early retirements of the F-22, left 'a lot riding on' the NGAD program.[52] This made the potential fallout far greater should the program face even half the delays, defects and other performance issues that the F-35 had, leaving the U.S. at risk of ceding a significant air superiority advantage to China. With the Air Force having given up on plans to expand its fighter fleet, conceded to repeated failures in its effort to raise operational rates,[53] deeply cut the F-22 program, and begun to seriously consider making deep cuts to F-35 orders, the NGAD may well have represented its last chance to reverse the trend towards relative decline in its position.

Beyond a next-generation fighter, the J-20 program was seen as a major factor spurring a wide range of other investments by the Pentagon. A number of investments modernising avionics of existing fighters were widely considered to be responses to the J-20's emergence, with the most notable examples being infrared sensors on the Navy's F-18E/F fighters[54] and the development of the new AN/APG-85 radar for the F-35. The new radar's primary developer, Northrop Grumman, described it as key to 'help ensure air superiority.'[55] In January 2022 upgrades for a portion of the F-22 fleet were also announced, and included belated IRST integration

U.S. Air Force E-3 Sentry AEW&C system. (Wikimedia Commons/Trevor McBride)

Royal Australian Air Force E-7A Wedgetail AEW&C system. (Wikimedia Commons/Royal Australian Air Force)

and upgrading of avionics to facilitate better sensor fusion.[56] A lack of both had seriously limited its situational awareness particularly against stealth targets compared to the F-35 or J-20, although the Raptor's older radar and lack of helmet-mounted sights, distributed aperture systems and other key features still left it a way behind.[57]

In March 2022, when confirming the first encounter between F-35s and J-20s, PACAF Commander General Kenneth Wilsbach stated that E-3 Sentry AEW&Cs in the region were suffering significant obsolescence issues, while strongly implying that as a result during the encounter 'our early warning aircraft could not see the J-20.' 'Those sensors that we rely on on the E-3 aren't really capable in the twenty-first century especially against a [stealth] platform like the J-20 or something similar to that. It just can't see those platforms far enough out to be able to provide an advantage to the shooters,' Wilsbach elaborated, stressing that 'that's why I would like to have the E-7.' This was aside from maintenance issues with the ageing aircraft which he emphasised seriously affected availability rates and often meant PACAF had no active AEW&C capability at

U.S. Air Force F-35 fires AIM-120 air-to-air missile. (U.S. Air Force/Christopher Okula)

and successful evolution and position as a leading contender for the title of the world's top fighter reflected the security and geopolitical trends of its time. The J-20A that entered production from 2019 was the first fighter to meet all operational criteria for the fifth generation. Where the F-35 could not supercruise, the F-22 could not fire high-off-boresight shots, use infrared sensors or share data properly as part of a broader network, and the Russian Su-57's stealth capabilities were relatively limited and its supercruise capability uncertain, the J-20A, J-20S and J-20B had none of these drawbacks.

The PLAAF's fielding of a large fleet of heavyweight fifth generation fighters was expected to have significant strategic implications for China's relations with its neighbours and particularly with the United States. A USAF presentation in 2009 stated that: 'Fifth generation fighters like the F-22A and the F-35 are key elements of our Nation's defence and ability for deterrence. As long as hostile nations recognise that U.S. airpower can strike their vital centres with impunity, all other U.S. Government efforts are enhanced, which reduces the need for military confrontation.'[63] The J-20 program not only ended this American advantage, but alongside parallel similarly rapid capability advances across the PLA it also increased China's ability to deter provocations or interventions in hotspots in the Taiwan Strait, Korea and the South and East China seas. While Lockheed Martin Aeronautics President James A. Blackwell contended in 1997 that the firm's new fighter was so capable that 'the first thing the F-22 will kill is the enemy's appetite for war,' the J-20 could well do much the same across multiple key hotspots.[64]

The prestige of being in a league of its own with the United States in developing one of the most complex and iconic kinds of weapons platforms, with the first full strength J-20 unit assembled an estimated seven years before any third country's fifth generation fighter would achieve the same, was a further not-inconsiderable benefit. One of the many manifestations of this was in popular media, and much like the F-22 was celebrated with considerable screen time in major pictures of the late 2000s, from *Iron Man* to *Transformers*, so too was the J-20 made an icon in Chinese films of the late 2010s and early 2020s such as *Sky Hunter* and *Wandering Earth II*. The fighter was even made the heart of its own film in 2023, *Born to Fly*, receiving treatment widely compared to that Hollywood had given America's most iconic and capable Cold War-era fighter – the F-14 – with *Top Gun* in 1986.

The development of Chinese military soft power and cultivation of the PLA's image as a world-leading force had implications beyond China's borders, with the J-20 being its most potent and recognisable icon. Indeed, as early as 2011 Taiwanese experts predicted that the J-20's success, alongside development of an array of capable support platforms such as the KJ-500, would transform the Asia Pacific's political landscape, and that the appeal of China's military strength

all.[58] The J-20 thus played a key role in spurring investment by the USAF in acquiring a new generation of AEW&C systems, with the E-7 Wedgetail with far superior electronics and sensors selected within a year of Wilsbach's statement. Funding for the aircraft was an unexpected new addition to the USAF's budget, with cuts to other programs expected to be made to fund the acquisition. The relatively sudden nature of the decision was widely seen as an indication that it was made in response to a new perceived challenge faced by U.S. forces – namely J-20 operations.[59]

The development of a new radar-guided air-to-air missile in the United States under the AIM-260 program from 2017 was also widely attributed to the need to counter J-20, both with better optimisation to engage stealth targets, and with a longer range to restore an advantage over the PL-15. The apparent superiority of the J-20's primary armament was met with alarm by U.S. officials, with the head of the USAF Air Combat Command Herbert Carlisle remarking in September 2017: 'Look at our adversaries and what they're developing, things like the PL-15 and the range of that weapon… How do we counter that and what are we going to do to continue to meet that threat?' He emphasised in an interview the following day: 'the PL-15 and the range of that missile, we've got to be able to out-stick that missile.'[60] Experts at *Aviation Week* described the PL-15 as having 'quickly provoked the U.S. Air Force to launch a new air-to-air missile program for the for the first time since the 1970s' in reference to the AIM-260 missile program which officials directly referred to as a response to the PL-15.[61] The missile was expected to integrate an AESA radar for inertial guidance much as the PL-15 had, with official Air Force statements highlighting that it was 'needed to remain ahead of adversary air threat investments.'[62] The missile was intended to complement, not replace, the older AIM-120 likely due to the unaffordability of acquisitions in comparable numbers.

THE J-20 IN THE EYES OF THE WORLD

Much as the F-22 was seen as the most widely recognised and potent a symbol of American military primacy, entering service as it did in 2005 with no rival program even at prototype stage, the J-20's fast would significantly increase the attractiveness of partnerships with Beijing in neighbouring countries.[65] The J-20 is thus expected to continue to play a central role in transforming narratives surrounding Chinese military capabilities not only among potential adversaries, but also among potential partners.

Following its unveiling the J-20 was widely disparaged in Western commentaries, and with the Western world very widely described by experts as being in denial regarding China's emerging technological and economic parity, responses to the fighter's emergence aligned with broader trends in how major Chinese achievements were responded to.[66] While as the first non-American stealth fighter the J-20's commonalities with the F-22 and F-35 were quicky highlighted by a wide range of Western sources to claim it had to be based on illicitly obtained American technologies, however, these claims were widely questioned by experts. The universal scientific rules of stealth technologies meant there were only so many ways to shape a fighter to benefit from a high kinematic performance while minimising its RCS – a very finite combination of edge alignments, chines, serrations, canted control surfaces, blended wing configurations, and intake geometry. With this being considered, the J-20's appearance was more distinct from American aircraft than the stealth fighters designed by any other country.

Entrenched but increasingly outdated perceptions in the West of Chinese high-tech, which drew on decades old stereotypes of both East Asian and communist states, appeared to strongly influence many assessments of the J-20's capabilities resulting in the aircraft consistently being underestimated. The rapid growth in the fighter's capabilities and those of its surrounding network, and the expected capabilities of future Chinese fighters which took many of the same technologies further, may well thus come as a very nasty surprise to Chinese adversaries when tested in combat. This would not be without precedent, with Chinese high-tech having achieved one 'Sputnik moment' after another from the early 2010s in areas ranging from 5G and quantum technologies to hypersonic glide vehicles to the consistent surprise of Western analysts.

The J-20's direct and most thoroughly combat-tested predecessor in the role of the PLAAF's top air superiority fighter – the MiG-15 – notably set a precedent in seriously challenging, and eventually shattering, the prevailing metanarrative in the West surrounding the industry which produced it. As USAF Lieutenant Colonel Earl J. McGill observed regarding the consensus before the MiG saw combat in Korea:

> Looking back, it seems remarkable that apparently no one in the West took the MiG-15 seriously... Although it made its world debut at the Tushino Air Show, the initial response was to largely dismiss the airplane as sub-standard. Even its NATO designation, Fagot (a worthless bundle of sticks), reflected the West's initial attitude toward the jet fighter. However, when the MiG-15 appeared in the skies over Korea, disinterest turned to shock and awe.[67]

Aviation Age was among the sources to highlight that the MiG-15's performance was evidence that Soviet aircraft designers were 'the exact antithesis to what most Americans would expect in a Communist state,'[68] with American journalist I. F. Stone highlighting 'growing awareness' among industry experts 'that the planes encountered in the skies over Korea were not mere models copied or stolen from the West by designers sweating it out in fear of the secret police.'[69] The extent to which the J-20 and China's defence sector were being disparaged, and the multiple indicators of a very high performance, mean the fighter could well have a similar impact should it ever see combat. The rate at which successive J-20 variants have improved, considerably faster than fighters in rival programs abroad have, means it is likely that the later this occurs, the greater the Chinese fighter's standing against rivals will be.

A NOTE ON WRITING AND RESEARCHING THIS BOOK, AND ON SOURCES OF REFERENCE USED

The information in this book compiles sources found over close to a decade of research efforts into key trends in aerial warfare and in combat aviation more specifically. Existing sources on the J-20 are generally restricted to releases by defence ministries and other government bodies, statements by both Chinese and American officials and personnel, and images either officially published by state affiliated outlets or, particularly in the program's earlier stages, leaked on the internet by Chinese aviation enthusiasts. These sources have generally focused on a single event in the fighter's development or service or a single aspect of the technologies that went into the aircraft, or else provided useful indications of how the J-20 and related programs are perceived by officials or personnel either in China or abroad. As a result, it is rare that any source is cited more than one or two times. Rather than presenting a separate bibliography and thus duplicating much material, the author therefore refers the reader to the extensive endnotes in this work.

Writing on the J-20 has posed some comparable albeit less extreme challenges to Cold War-era efforts to assess the development of leading non-exported Soviet fighters and interceptors, whether the Su-15 in the 1960s or the MiG-31 and Su-27 in the 1980s. Despite the onset of the information age, and China's much greater openness by Soviet standards regarding its defence programs, Chinese fighter programs still maintain significantly greater opacity surrounding them than those of any other country. As a result, much of what has been written on the J-20 particularly outside China has been highly speculative, with parts of this book tracing changes in prevailing views towards the fighter abroad as more information has become available, and assessing the underlying reasons why various conclusions were reached by analysts at various times in the aircraft's development. A range of general and specialist media and think tank sources have provided useful means of following these changes.

As the body of available information on the J-20 program has grown, a more comprehensive understanding has increasingly become possible particularly when this information is put in the context of both broader trends in Chinese industry and high-tech and of China's threat perceptions and prioritised capabilities. The writer has thus referred to a very wide range of sources as part of a qualitative analysis of primarily text, photo and video evidence, including a high reliance on interviews conducted by third parties.

NOTES

Chapter 1

1. Li, Xiaobing, *China's Battle for Korea: The 1951 Spring Offensive* (Bloomington: Indiana University Press, 2014), p. 11. Hanley, Charles J., *Ghost in Flames: Life & Death in a Hidden War, Korea 1950-53* (New York: Public Affairs, 2020), p. 31.
2. Epstein, Israel, *My China Eye: Memoirs of a Jew and a Journalist* (San Francisco: Long River Press, 2005), p. 251. Bodenheimer, Thomas and Gould, Robert, *Rollback!: Right-wing Power in U.S. Foreign Policy* (Boston: South End, 1989), p. 18. Philips, Steve, *The Cold War: Conflict in Europe and Asia* (Oxford, Heinemann, 2001), pp. 71–72. Spurr, Russel, *Enter the Dragon: China's Undeclared War Against the U.S. in Korea, 1950-1951* (New York: William Morrow, 2010), p. 106.
3. Salisbury, Harrison E., *The New Emperors: China in the Era of Mao and Deng* (New York: Avon Books, 1992), p. 106.
4. Testimony of Dean Acheson, Hearings held in executive session before the U.S. Senate Foreign Relations Committee during 1949-1950, p. 23. Truman, Harry S., *Memoirs, Volume Two: Years of Trial and Hope, 1946-1953* (New York: Doubleday & Company, 1956), p. 66. Harris Smith, Richard, *OSS: The Secret History of America's First Central Intelligence Agency* (Berkley: University of California Press, 1972), pp. 259–282. Fleming, Denna Frank, *The Cold War and its Origins, 1917-1960* (Crows Nest: Allen and Unwin, 1961), p. 570. Tuchman, Barbra W., *Sitwell and the American Experience in China 1911-1945* (London: MacMillan, 1970), pp. 666–677. 'Letter to Congressman Hugh de Lacy of State of Washington,' *Congressional Record*, January 24, 1946, Appendix, vol. 92, part 9, p. A225.
5. Appleman, Roy E., *South to the Naktong, North to the Yalu: United States Army in the Korean War* (Washington DC: Department of the Army, 1998), p. 755.
6. Hanley, Charles J., *Ghost in Flames: Life & Death in a Hidden War, Korea 1950-53* (New York: Public Affairs, 2020), pp. 144–145.
7. Li, Xiaobeing, *China's Battle for Korea: The 1951 Spring Offensive* (Bloomington: Indiana University Press, 2014), p. 68. Futrell, Robert F., *The United States Air Force in Korea 1950-1953* (Washington DC, Office of Air Force History, 1983), p. 339. Stokesbury, James L., *A Short History of the Korean War* (New York: William Morrow, 1998), pp. 55–56.
8. Hanley, Charles J., *Ghost in Flames: Life & Death in a Hidden War, Korea 1950-53*, New York, Public Affairs, 2020, p. 267.
9. Hanley, Charles J., *Ghost in Flames: Life & Death in a Hidden War, Korea 1950-53*, (New York: Public Affairs, 2020), p. 268.
10. Li, Xiaobeing, *China's Battle for Korea: The 1951 Spring* Offensive (Bloomington: Indiana University Press, 2014), pp. 61, 69.
11. Futrell, Robert Frank, *The United States Air Force in Korea, 1950-1953* (Washington DC: Office of Air Force History, 1983), pp. 689–690. Williams, William J., *A Revolutionary War: Korea and the Transformation of the Postwar World* (Chicago: Imprint, 1993), pp. 144–145. Shen, Zonghong et al., 中国人民志愿军抗美援朝战士, [*A History of the War to Resist America and Assist Korea by the Chinese People's Volunteers*] (Beijing: Military Science Press, 1988), p. 319.
12. Li, Xiaobeing, *China's Battle for Korea: The 1951 Spring* Offensive (Bloomington: Indiana University Press, 2014), p. 121. Hanley, Charles J., *Ghost in Flames: Life & Death in a Hidden War, Korea 1950-53* (New York: Public Affairs, 2020), pp. 269, 280.
13. Hanley, Charles J., *Ghost in Flames: Life & Death in a Hidden War, Korea 1950-53* (New York: Public Affairs, 2020), p. 421.
14. Hanley, Charles J., *Ghost in Flames: Life & Death in a Hidden War, Korea 1950-53* (New York: Public Affairs, 2020), p. 269.
15. Xu, Yan, 第一次较量: 抗美援朝战争的历史回顾与反思 [*The First Test of Strength: A Historical Review and Evaluation of the War to Resist America and Assist Korea*] (Beijing: Chinese Broadcasting and Television Press, 1990), p. 207. Qi, Dexue et al., 'Enlightenment on Defeating a Strong Enemy with Inferior Weapons and Equipment in the Korean War,' 军事历史 [*Military History*], no. 6, 1999, p. 33.
16. Stone, I. F., *Hidden History of the Korean War*, Boston (Boston: Little, Brown and Company, 1988), p. 341.
17. McGill, Earl J., *Black Tuesday Over Namsi: B-29s vs MiGs – The Forgotten Air Battle of the Korean War* (Solihull: Helion & Company, 2012), Chapter 4: The Machinery of War.
18. McGill, Earl J., *Black Tuesday Over Namsi: B-29s vs MiGs – The Forgotten Air Battle of the Korean War* (Solihull: Helion & Company, 2012), Chapter 9: Analysis, Conclusions and Reflections.
19. Hanley, Charles J., *Ghost in Flames: Life & Death in a Hidden War, Korea 1950-53* (New York: Public Affairs, 2020), p. 358.
20. Appleman, Roy E., *Escaping the Trap: The U.S. Army X Corps in Northeast Korea, 1950* (College Station: A&M University Press, 1990), pp. 367–368. Hastings, Max, *Korean War* (London: Michael Joseph, 1988), p. 170.
21. Chan, Minnie, 'China's air force turns 70 with tales of its dare-to-die Korean war pilots,' *South China Morning Post*, November 10, 2019.
22. Regarding the discrepancies in experience with the adversary, Korean War veteran Chinese pilot Li Han highlighted in his memoirs: 'We had a strong spirit even though our fighting experience was weak. We tried to engage them in combat but we knew that if we couldn't keep fighting, then we should just crash into them.' Another PLAAF veteran recalled: 'Almost all of the Chinese pilots in the Korean war were "dare to die"… their spirits were so strong they were ready to die.' 'China's air force turns 70 with tales of its dare-to-die Korean war pilots,' *South China Morning Post*, November 10, 2019.
23. Appleman, Roy E., *Escaping the Trap: The U.S. Army X Corps in Northeast Korea, 1950* (College Station: A&M University Press, 1990), p. 375. Zhang, Xiaoming, *Red Wings over the Yalu: China, the Soviet Union, and the Air War in Korea* (College Station: Texas A&M University Press, 2003), p. 132.
24. Joiner, Stephen, 'The Jet that Shocked the West, How the MiG-15 grounded the U.S. bomber fleet in Korea,' *Air & Space Magazine*, December 2013.
25. Stone, I. F., *Hidden History of the Korean War* (Boston: Little, Brown and Company, 1988), p. 339.
26. Nash, Chris, *What is Journalism? The Art and Politics of a Rupture* (London: Palgrave Macmillan, 2016), p. 91.
27. *New York Times*, January 20, 1952.
28. *Aviation Week*, December 17, 1951. *New York Dally Compass*, January 3, 1952.
29. Stone, I. F., *Hidden History of the Korean War* (Boston: Little, Brown and Company, 1988), p. 434.
30. Hanley, Charles J., *Ghost in Flames: Life & Death in a Hidden War, Korea 1950-53* (New York: Public Affairs, 2020), p. 359. Associated Press, February 14, 1952.
31. *New York Times*, January 20, 1952.
32. Hanley, Charles J., *Ghost in Flames: Life & Death in a Hidden War, Korea 1950-53* (New York: Public Affairs, 2020), p. 348.
33. Stone, I. F., *Hidden History of the Korean War* (Boston: Little, Brown and Company, 1988), p. 344.
34. Stone, I. F., *Hidden History of the Korean War* (Boston: Little, Brown and Company, 1988), p. 344.
35. *New York Times*, December 9, 1951. Stone, I. F., *Hidden History of the Korean War* (Boston: Little, Brown and Company, 1988), p. 344.
36. *Washington Post*, May 22, 1952.

37. Bruning Jr., John R., *Crimson Sky: The Air Battle for Korea* (Dulles: Brassey's, 1999), p. xiv. Crane, Conrad C., *American Airpower Strategy in Korea, 1950-1953* (Lawrence: University Press of Kansas, 2000), p. 107.
38. Bruning Jr., John R., *Crimson Sky: The Air Battle for Korea* (Dulles: Brassey's, 1999), p. xiv.
39. Zhang, Xiaoming, *Red Wings over the Yalu: China, the Soviet Union, and the Air War in Korea* (College Station: Texas A&M University Press, 2003), p. 201.
40. Wang Dinglie et al., 当代中国空军 [*China's Current Air Force*] (Beijing: China Social Science Press, 1989), pp. 208, 209.
41. Wang Dinglie et al., 当代中国空军 [*China's Current Air Force*] (Beijing: China Social Science Press, 1989), p. 210.
42. Wang Dinglie et al., 当代中国空军 [*China's Current Air Force*] (Beijing: China Social Science Press, 1989), p. 210.
43. Zhang, Lin and Ma, Changzhi, 中国元帅徐向前 [*Chinese Marshal Xu Xiangqian*] (Beijing: Communist Party of China Central Academy Press, 1995), p. 404.
44. Hastings, Max, *Korean War* (London: Michael Joseph, 1988), p. 170.
45. Ham, Paul, *Hiroshima Nagasaki: The Real Story of the Atomic Bombings and their Aftermath* (New York: Doubleday, 2012), pp. 488–490.
46. Harvey, Ian, 'FBI Files Reveals Winston Churchill's Secret Bid to Nuke Russia to Win Cold War,' *War History Online*, November 17, 2014.
47. Acheson, Dean G., *Present at the Creation: My Years in the State Department* (London: W. W. Norton, 1969), pp. 463–464, Far Eastern Air Forces HQ to MacArthur, 8 November 1950, RG 6 Far East Command, Box 1, General Files 10, Correspondence Nov-Dec 1950, MacArthur Memorial Library, Norfolk VA.
48. Jackson, Robert, *Air War Over Korea* (London: Ian Allen, 1973), p. 61.
49. Harden, Blaine, 'The U.S. war crime North Korea won't forget,' *Washington Post*, March 24, 2015.
50. Thames Television, transcript for the fifth seminar for: *Korea: The Unknown War*, November 1986.
51. LeMay, Curtis and Cantor, MacKinley, *Mission with LeMay* (New York: Doubleday, 1965), p. 382.
52. Truman, Margaret, *Harry S. Truman* (New York: William Morrow & Company, 1973), pp. 495–496, 'USE OF A-BOMB IN KOREA STUDIED BY U.S.—TRUMAN,' *Pittsburgh Press*, November 30, 1950, p. 1. The President's New Conference, November 30, 1950, The American Presidency Project, University of California at Santa Barbara.
53. 'Thaw in the Koreas?,' *Bulletin of Atomic Scientists*, vol. 48, no. 3, April 1992, pp. 18, 19. Weintraub, Stanley, *MacArthur's War: Korea and the Undoing of an American Hero* (New York: Free Press, 2000), p. 263. Millet, Alan R., *Their War for Korea* (Washington DC: Brassey's, 2002), p. 169.
54. Grosscup, Beau, *Strategic Terror: The Politics and Ethics of Aerial Bombardment* (London: Zed Books, 2013), Chapter 5: Cold War Strategic Bombing: From Korea to Vietnam, Part 4: The Bombing of Korea. Levine, Alan J., *Stalin's Last War; Korea and the Approach to World War III* (Jefferson: McFarland & Company, 2005), pp. 278, 280.
55. Pape, Robert A., *Bombing to Win; Air Power and Coercion in War* (Ithaca: Cornell University Press, 1996), p. 146.
56. Wang, Yan, 彭德怀转 [*Biography of Peng Dehuai*] (Beijing: Today's China Press, 1993), pp. 502, 506.
57. G-3 381 Pacific, G-3 Staff Study, 'Capability of U.S. Army to Implement Commander in Chief, United Nations Command Operations Plan,' ca. 21, Jan 53. Levine, Alan J., *Stalin's Last War; Korea and the Approach to World War III* (Jefferson: McFarland & Company, 2005), pp. 277, 278, 280. BBC Summary, Far East, No. 221, January 23, 1953. Chang, Su-Ya, *Unleashing Chiang Kai-shek? Eisenhower and the Policy of Indecision toward Taiwan, 1953* (Taipei: Institute of Modern History, 1991), Kuo, Fang Pu and Shih, Cheng Chu, *The Working Record of the U.S. Military Assistance Advisory Group: The Headquarters* (Taipei: Historic Office, Republic of China Ministry of National Defence, 1981), pp. 10–12. Hanley, Charles J., *Ghost in Flames: Lift & Death in a Hidden War, Korea 1950-53* (New York: Public Affairs, 2020), p. 257.
58. Journal of the American Intelligence Professional, unclassified articles from: *Studies in Intelligence*, vol. 57, no. 3, September 2013, pp. 22–28.
59. 'Two CIA Prisoners in China, 1952–1973,' Website of the Central Intelligence Agency, News & Information, April 5, 2007.
60. 'Idaho Crash Reveals Secret U-2 Training of Chinese,' *New York Times,* August 30, 1964. 'Pilotless U.S. Plane Downed, China Says,' *New York Times,* November 17, 1964. Journal of the American Intelligence Professional, unclassified articles from: *Studies in Intelligence*, vol. 57, no. 3, September 2013, pp. 22–28.
61. Journal of the American Intelligence Professional, unclassified articles from: *Studies in Intelligence*, vol. 57, no. 3, September 2013, p. 28.
62. Journal of the American Intelligence Professional, unclassified articles from: *Studies in Intelligence*, vol. 57, no. 3, September 2013, p. 28.
63. Rupprecht, Andreas, *Modern Chinese Warplanes: Chinese Air Force – Comat Aircraft and Units* (Houston: Harpia, 2018), p. 17.
64. Kirby, William C., 'The Internationalization of China: Foreign Relations at Home and Abroad in the Republican Era,' *China Quarterly*, no. 150, June 1997, p. 458.
65. Didly, Douglas C. and Thompson, Warren E., *F-86 Sabre vs MiG-15: Korea 1950-1953* (Oxford: Osprey, 2013), p. 77.
66. Gao, Charlie, 'In 1971, Indian Mig-21's Beat Some American-Built F-104A Starfighters. Was It a Fluke?,' *National Interest,* August 5, 2018.
67. Buttler, Tony and Gordon, Yefim, *Soviet Secret Projects: Fighters Since 1945* (Hinckley: Midland Publishing, 2005), pp. 75–77, 80, 81, 92, 107. Mason, R. A. and Taylor, John W. R., *Aircraft, Strategy and Operations of the Soviet Air Force* (London, Jane's, 1986), pp. 66, 134, 137. 'How Soviet Leader Nikita Khrushchev Let America Win the Race to Develop the Best Fighter Jets,' *Military Watch Magazine*, February 9, 2021. Lambeth, Benjamin S., *Russia's Air Power at the Crossroads* (Santa Monica: RAND, 1996), pp. 33–34. Grinyuk, Dmitri and Butowski, Piotr, 'An Unusual Conversation at the Main Staff,' *Krylia Rodiny,* no. 11, November 1991.
68. 'How Soviet Leader Nikita Khrushchev Let America Win the Race to Develop the Best Fighter Jets,' *Military Watch Magazine,* February 9, 2021.
69. Lake, Jon, *Jane's How to Fly and Fight in the Mikoyan MiG-29 Fulcrum* (New York City: Harper Collins, 1998), Introduction. Lake, Jon, *MiG-29: Soviet Superfighter* (London, Osprey, 1989), p. 7.
70. Gordon, Yefim, *Sukhoi Su-27* (Hinckley: Midland Publishing, 2007), p. 13. Farley, Robert, 'Why The F-15 Must Fear Russia's Su-27 Fighter,' *National Interest,* February 1, 2020. Gordon, Yefim and Komissarov, Dmitry, *Sukhoi Su-57* (Manchester: Hikoki Publications, 2021), p. 7.
71. Gordon, Yefim, *Sukhoi Su-27* (Hinckley: Midland Publishing, 2007), pp. 514–515.
72. Gordon, Yefim, *Sukhoi Su-27* (Hinckley: Midland Publishing, 2007), pp. 515–516.
73. Gordon, Yefim, *Sukhoi Su-27* (Hinckley: Midland Publishing, 2007), p. 514.
74. *Su-27 for DCS World* (Moscow: Eagle Dynamics, 2014), p. 3.
75. Lake, Jon, *Su-27 Flanker: Sukhoi Superfighter* (London: Osprey, 1992), p. 88. Gordon, Yefim, *Sukhoi Su-27* (Hinckley: Midland Publishing, 2007), p. 522.
76. Gordon, Yefim, *Sukhoi Su-27* (Hinckley: Midland Publishing, 2007), pp. 6, 525. *Su-27 for DCS World* (Moscow: Eagle Dynamics, 2014), p. 3. Lake, Jon, *Su-27 Flanker: Sukhoi Superfighter* (London: Osprey, 1992), p. 111.

77. Gordon, Yefim, *Sukhoi Su-27* (Hinckley: Midland Publishing, 2007), p. 25, 517.
78. Gordon, Yefim, *Sukhoi Su-27* (Hinckley: Midland Publishing, 2007), p. 519.
79. Gordon, Yefim and Davidson, Peter, *Sukhoi Su-27 Flanker* (North Branch: Specialty Press, 2006), p. 18. Gordon, Yefim, *Sukhoi Su-27* (Hinckley: Midland Publishing, 2007), pp. 517–518.
80. Lake, Jon, *Su-27 Flanker: Sukhoi Superfighter* (London: Osprey, 1992), Introduction.
81. Gordon, Yefim and Davidson, Peter, *Sukhoi Su-27 Flanker* (North Branch: Specialty Press, 2006), p. 22.
82. Lilley, James and Shambaugh, David L., *China's Military Faces the Future* (Abingdon: Routledge, 2015), pp. 96–99. Lake, Jon, *Su-27 Flanker: Sukhoi Superfighter* (London: Osprey, 1992), Introduction. Ilin, Vladimir, 'Air Bases of Russia: Lipetssk – One of the Aviation Centres of Russia,' Vestnik Vozdushnogo Flota, March 14, 1995, in: Foreign Broadcast Information Service Military Affairs 95-148-S, March 14, 1995.
83. Gordon, Yefim, *Sukhoi Su-27* (Hinckley: Midland Publishing, 2007), p. 524.
84. *Department of Defense Appropriations for 2002: Hearings Before a Subcommittee of the Committee on Appropriations House of Representatives*, One Hundred and Seventh Congress, First Session, Subcommittee on Defense, Washington DC, U.S. Government Printing Office, 2004, p. 813.
85. Lake, Jon, *Su-27 Flanker: Sukhoi Superfighter* (London: Osprey, 1992), Introduction.
86. Gordon, Yefim, *Sukhoi Su-27* (Hinckley: Midland Publishing, 2007), pp. 522–523.
87. Gordon, Yefim, *Sukhoi Su-27* (Hinckley: Midland Publishing, 2007), p. 522.
88. Gordon, Yefim, *Sukhoi Su-27* (Hinckley: Midland Publishing, 2007), pp. 533–534.
89. Gordon, Yefim, *Sukhoi Su-27* (Hinckley: Midland Publishing, 2007), pp. 531–534. Baldauf, Scott, 'Indian Air Force, in war games, gives US a run,' *Christian Science Monitor*, November 28, 2005.
90. Press, Daryl G., 'The Myth of Air Power in the Persian Gulf War and the Future of Warfare,' *International Security*, vol. 26, no. 2, October 2001, pp. 5–44.
91. Khalilzad, Zalmay and Shapiro, Jeremy, *The United States Air and Space Power in the 21st Century* (Santa Monica: RAND, 2002), Chapter Three.
92. Gons, Eric Stephen, *Access Challenges and Implications for Airpower in the Western Pacific* (Santa Monica: RAND, 2011), pp. iii, 1.
93. 解放军报 (*Jiefangjun Bao*), January 25, 1991.
94. Biddle, Stephen, *Military Power: Explaining Victory and Defeat in Modern Battle* (Princeton: Princeton University Press, 2006), p. 20. Cordesman, Anthony H., 'The Real Revolution in Military Affairs,' *Centre for Strategic and International Studies,* August 5, 2014.
95. 'Air-Sea Battle Doctrine: A Discussion with the Chief of Staff of the Sir Force and Chief of Naval Operations,' *Brookings Institute,* May 16, 2012.
96. 'Flashback: 1991 Gulf War,' *BBC News*, March 20, 2003. 'Sanctions Blamed for Deaths of Children,' *Lewiston Morning Tribune*, December 2, 1995. Stahl, Lesley, 'Interview with Madeline Albright,' *60 Minutes*, May 12, 1996.
97. *The Mirror* (Hong Kong), January 22, 1991. Lam, Willy Wo-Lap, 'Iraq war hands lessons to China,' *CNN*, April 15, 2003.
98. Erickson, Amanda, 'The last time the U.S. was on "the brink of war" with North Korea,' *Washington Post,* August 9, 2017. Cumings, Bruce, *Korea's Place in the Sun* (New York: W. W. Norton and Company, 1997), p. 428.
99. Schmitt, Eric, 'In a Fatal Error, C.I.A. Picked a Bombing Target Only Once: The Chinese Embassy,' *New York Times*, July 23, 1999. 'Truth behind America's raid on Belgrade,' *The Guardian,* November 28, 1999. Sweeney, John and Holsoe, Jens and Vulliamy, Ed, 'Nato bombed Chinese deliberately,' *The Guardian,* October 17, 1999.
100. 'An Economic Analysis of the Changes in China's Military Expenditure in the Last Ten Years,' *Jingji Yanjiu* [Economic Studies], no. 6, June 20, 1990, pp. 77–81. Jencks, Harlan W., 'Chinese Evaluations of "Desert Storm": Implications for PRC Security,' *The Journal of East Asian Affairs*, vol. 6, no. 2, Summer/Fall 1992, p. 462.
101. Foreign Broadcast Information Service Daily Report-China 90- 153, pp. 36–38. Jencks, Harlan W., 'Chinese Evaluations of "Desert Storm": Implications for PRC Security,' *The Journal of East Asian Affairs*, vol. 6, no. 2, Summer/Fall 1992, p. 463.
102. *South China Morning Post*, March 25, 1991, pp. 1, 11.
103. *Jiefangjun Bao*, March 28, 1991.
104. *Hong Kong Standard,* April 16, 1991. *Beijing Radio*, April 16, 1991. Jencks, Harlan W., 'Chinese Evaluations of "Desert Storm": Implications for PRC Security,' *The Journal of East Asian Affairs*, vol. 6, no. 2, Summer/Fall 1992, p. 465.
105. Tyson, James L., 'Workers of the World, Compute: China Revamps Industry,' *Christian Science Monitor*, July 16, 1991, p. 4.
106. Jencks, Harlan W., 'Chinese Evaluations of "Desert Storm": Implications for PRC Security,' *The Journal of East Asian Affairs*, vol. 6, no. 2, Summer/Fall 1992, p. 463–465. Foreign Broadcast Information Service Daily Report-China, 91–041, pp. 33–35. Luo Xiaobing, 'Strengthening National Defense Building Is Important Guarantee for Economic Development…,' *Jiefangjun Bao,* February 6, 1991.
107. *Xinhua*, March 16, 1991. Foreign Broadcast Information Service Daily Report-China 91–054, pp. 27–28. Jencks, Harlan W., 'Chinese Evaluations of "Desert Storm": Implications for PRC Security,' *The Journal of East Asian Affairs*, vol. 6, no. 2, Summer/Fall 1992, p. 463.
108. International Institute for Strategic Studies, *The Military Balance*, Volume 103, 2003, p. 115.
109. 'Aerospace Nation: Gen Kenneth S. Wilsbach,' *Mitchell Institute for Aerospace Studies* (YouTube Channel), March 15, 2022.
110. Wezeman, Siemon T., 'China, Russia and the shifting landscape of arms sales,' *Stockholm International Peace Research Institute,* July 5, 2017. Menshikov, Stanislav, 'Russian Capitalism Today,' *Monthly Review*, vol. 51. no. 3, 1999, pp. 82–86. Millar, James R., 'Can Putin Jump-Start Russia's Stalled Economy?,' *Current History*, vol. 99, no. 639, October 2000, pp. 329–333.
111. Wezeman, Siemon T., 'China, Russia and the shifting landscape of arms sales,' *Stockholm International Peace Research Institute,* July 5, 2017. Cooper, Julian, 'The Future of the Russian Defense Industry' in: Allison, Roy and Bluth, Christopher, *Security Dilemmas in Russia and Eurasia* (London: Royal Institute for International Affairs, 1998), p. 96.
112. Gordon, Yefim, *Sukhoi S-37 and Mikoyan MFI: Russian Fifth-Generation Fighter Technology Demonstrators* (Hinckley: Midland Publishing, 2001), pp. 12, 13. Buttler, Tony and Gordon, Yefim, *Soviet Secret Projects: Fighters Since 1945* (Hinckley: Midland Publishing, 2005), pp. 153–155.
113. Butowski, Piotr, *Su-57 Felon* (Stamford: Key Publishing, 2022), 'Chapter 1: The First Approach: MiG MFI.'
114. 'Air Combat with "Ghosts",' *Aviatsiia i kosmonavtika,* no. 7, July 1991, pp. 4, 5. Lambeth, Benjamin S., *Russia's Air Power at the Crossroads* (Santa Monica: RAND, 1996), p. 240.
115. Gordon, Yefim, *Sukhoi S-37 and Mikoyan MFI: Russian Fifth-Generation Fighter Technology Demonstrators* (Hinckley: Midland Publishing, 2001), pp. 21, 22.
116. Gordon, Yefim, *Sukhoi S-37 and Mikoyan MFI: Russian Fifth-Generation Fighter Technology Demonstrators* (Hinckley: Midland Publishing, 2001), pp. 12–13, 15, 31.
117. Gordon, Yefim, *Sukhoi S-37 and Mikoyan MFI: Russian Fifth-Generation Fighter Technology Demonstrators* (Hinckley: Midland Publishing, 2001), p. 22.
118. Gordon, Yefim and Komissarov, Dmitry, *Sukhoi Su-57* (Manchester: Hikoki Publications, 2021), p. 25.

119. Gordon, Yefim and Komissarov, Dmitry, *Sukhoi Su-57* (Manchester: Hikoki Publications, 2021), p. 42. Buttler, Tony and Gordon, Yefim, *Soviet Secret Projects: Fighters Since 1945* (Hinckley: Midland Publishing, 2005), pp. 153–155. Gordon, Yefim, *Sukhoi S-37 and Mikoyan MFI: Russian Fifth-Generation Fighter Technology Demonstrators* (Hinckley: Midland Publishing, 2001), p. 35.
120. Had the MiG 1.42 program been just two years further ahead in development by 1991 it likely would have had the momentum needed to survive the Soviet Union's disintegration and enter service in the 2000s – providing a far superior aircraft to anything Russia would field at least until the mid-late 2020s. The fighter was referred to in the early 2000s by Russian Air Force representatives as 'not even tomorrow's aircraft but truly the day after tomorrow's aircraft' for its tremendous advances. As aviation expert Yefim Gordon summarised: 'in retrospect it is clear that the [MiG 1.42] program philosophy was better than that of the American ATF [Advanced Tactical Fighter]… it held promise of a breakthrough in aerodynamics, automation of flight control systems, ergonomics AI etc. And so it did, except that the Soviet aircraft industry did not have enough time for practical implementation of these results before the tempestuous events of 1990-91.' Gordon, Yefim and Komissarov, Dmitry, *Sukhoi Su-57* (Manchester: Hikoki Publications, 2021, pp. 10, 44.
121. Rupprecht, Andreas, *Modern Chinese Warplanes: Chinese Air Force – Comat Aircraft and Units* (Houston: Harpia, 2018), pp. 37, 38.
122. 'SIPRI Military Expenditure Database,' *Stockholm International Peace Research Institute* (https://milex.sipri.org/sipri).
123. Mañé Estrada, Aurèlia and de la Cámara Arilla, Carmen, 'Is Russia Drifting toward an Oil-Rentier Economy?,' *Eastern European Economics*, vol. 43, no. 5, September-October 2005, pp. 46–73. Wagstyl, Stefan, 'Russia: riding with the rentiers,' *Financial Times*, July 8, 2011.

Chapter 2

1. 'China and the world: Inside the dynamics of a changing relationship,' *McKinsey Global Institute*, July 1, 2019.
2. Allison, Graham, 'China Is Now the World's Largest Economy. We Shouldn't Be Shocked,' *National Interest,* October 15, 2020.
3. Gaida, Jamie et al., 'ASPI's Critical Technology Tracker: The global race for future power,' Australian Strategic Policy Institute, Policy Brief, Report No. 69, 2023.
4. 'Vital Signs 2020: The Health and Readiness of the Defense Industrial Base,' *National Defense Industrial Association*. Nebehay, Stephanie, '"Driving force" China accounts for nearly half global patent filings: U.N.,' *Reuters*, October 15, 2019.
5. Allison, Graham, 'The Great Rivalry: China vs. the U.S. in the 21st Century,' *Belfer Centre for Science and International Affairs, Harvard Kennedy School*, December 7, 2021.
6. 'Tim Cook Discusses Apple's Future in China,' *Fortune Magazine,* December 5, 2017. Nicas, Jack, 'A Tiny Screw Shows Why iPhones Won't Be "Assembled in U.S.A.",' *New York Times,* January 28, 2019. Duhig, Charles and Bradsher, Keith, 'How the U.S. Lost Out on iPhone Work,' *New York Times,* January 21, 2012.
7. Duhigg, Charles and Bradsher, Keith, 'How the U.S. Lost Out on iPhone Work,' *New York Times,* January 21, 2012.
8. Gould, Joe and Losey, Stephen, 'Amid hiring boom, defense firms say labor shortage is dragging them down,' *Defense News*, August 5, 2022. Tirpak, John A., 'F-35 Production Challenged to Keep Up with Demand,' *Air and Space Forces,* July 3, 2023. 'US defence contractors squeezed by shortages of labour and parts,' *Financial Times,* July 26, 2022. 'Parts shortages dog US defence contractors as war depletes arsenals,' *Financial Times,* October 27, 2022.
9. '杨伟代表：歼20将进行系列化发展 不断提升作战能力' ('Representative Yang Wei: The J-20 will undergo serialised development to continuously enhance its combat capabilities.'), *Renmin Wang*, March 20, 2018.
10. Tirpak, John A., 'Kendall: Digital Engineering Was "Over-Hyped," But Can Save 20 Percent on Time and Cost,' *Air and Space Forces*, May 23, 2023.
11. '"歼－２０，你不要飞得那么快……"' ('J-20, Don't Fly So Fast……'), *Xinhua*, November 10, 2018.
12. Erickson, Andrew S. and Lu, Hanlu and Bryan, Kathryn and Septembre, Samuel, 'Research, Development, and Acquisition in China's Aviation Industry: The J-10 Fighter and Pterodactyl UAV,' *Study of Innovation and Technology in China: University of California Institute of Global Conflict and Cooperation*, no. 8, 2014.
13. Freedberg Jr., Sydney J., 'Chinese Air Force Tries Hard But Plays Catch-Up With US; Watch PLA Espionage,' *Breaking Defense*, September 18, 2012.
14. 'China's J-20: future rival for air dominance?,' *Strategic Comments*, vol. 17, issue 1, 2011, pp. 1–3.
15. 'China's J-20: future rival for air dominance?,' *Strategic Comments*, vol. 17, issue 1, 2011, pp. 1–3.
16. 'China's J-20: future rival for air dominance?,' *Strategic Comments*, vol. 17, issue 1, 2011, pp. 1–3.
17. '最新：中国空军选定下一代战机由611所方案胜出' ('Latest: The Chinese Air Force selected the next generation of fighter jets by the 611 program to win'), 亚东军事网, November 5, 2010.
18. Chang, Yihong, 'China Launches New Stealth Fighter Project,' *Jane's Defence Weekly*, December 2002.
19. Knight, Will, 'Chinese Stealth Fighter Plans Revealed,' *New Scientist*, December 2002.
20. Rupprecht, Andreas, 'The PLA Air Force's "Silver-Bullet" Bomber Force,' *Jamestown Foundation China Brief,* Volume 17, Issue 10, July 21, 2017. Rupprecht, Andreas, *Modern Chinese Warplanes: Chinese Air Force – Comat Aircraft and Units*, Houston, Harpia, 2018, p. 40.
21. Davies, Peter E. and Thornborough, Tony, *McDonnell Douglas F-15 Eagle* (Ramsbury: Crowood Press, 2001), p. 164.
22. Niemi, Christopher J., 'The F-22 Acquisition Program: Consequences for the US Air Force's Fighter Fleet,' *Air & Space Power Journal,* vol. 26, no. 6, November-December 2012, p. 65.
23. Davies, Peter E. and Thornborough, Tony, *McDonnell Douglas F-15 Eagle* (Ramsbury: Crowood Press, 2001), p. 164.
24. Younossi, Obaid et al., *Lessons Learned from the F/A-22 and F/A-18E/F Development Programs* (Santa Monica: RAND, 2005), p. 4. Niemi, Christopher J., 'The F-22 Acquisition Program: Consequences for the US Air Force's Fighter Fleet,' *Air & Space Power Journal,* vol. 26, no. 6, November-December 2012, pp. 53–82.
25. O'Hanlon, Michael E., 'The Plane Truth: Fewer F-22s Mean a Stronger National Defense,' *Brookings Institute*, September 1, 1999. Battle Stations, Season 1, Episode 39, F-22 Raptor.
26. Wheeler, Winslow, 'What does an F-22 Cost?,' *Project on Government Oversight*, March 30, 2009.
27. For cost cutting cancellation of F-22's IRST and side facing radars see: Rogoway, Tyler, 'No, The Su-57 Isn't "Junk:" Six Features We like On Russia's New Fighter,' *The Drive,* April 20, 2018.
28. Vartabedian, Ralph and Hennigan, W. J., 'F-22 program produces few planes, soaring costs,' *Los Angeles Times*, June 16, 2013.
29. 'Documents show Air Force neglected concerns about F-22 pilot safety,' *Public Integrity,* September 27, 2012. Axe, David, 'US Stealth Jets Choking Pilots at Record Rates,' *Wired,* June 14, 2012. Axe, David, 'Air Force to Stealth Fighter Pilots: Get Used to Coughing Fits,' *Wired,* February 25, 2013. Fabey, Michael, 'USAF Deciphers "Mosaic" Of F-22 Oxygen Supply Problems,' *Aviation Week*, August 1, 2012. 'Air force pilots describe health problems from flying F-22 jet,' *CBS News* (YouTube Channel), May 7, 2012.
30. 'F-22 avionics designers rely on obsolescent electronics, but plan for future upgrades,' *Military Aerospace,* May 1, 2001. Everstine,

Brian W., 'The F-22 and the F-35 Are Struggling to Talk to Each Other … And to the Rest of USAF,' *Air Force Magazine*, January 29, 2018.

31. Lim, Louisa, 'China Uncloaks Stealth Fighter Prototype,' *NPR*, January 5, 2011. Rizzo, Jennifer and Keyes, Charley, 'Is China closer than thought to matching U.S. fighter jet prowess?,' *CNN*, January 6, 2011.
32. 'USAF: F-22 to Be Phased-Out Within Next Decade,' *The Defense Post*, May 21, 2021.
33. Rizzo, Jennifer and Keyes, Charley, 'Is China closer than thought to matching U.S. fighter jet prowess?,' *CNN*, January 6, 2011.
34. 'China stealth plane still "years away", says Pentagon,' *BBC News*, January 6, 2011.
35. Bodeen, Christopher, 'China's stealth fighter photos cause an international stir,' *NBC News*, January 6, 2011.
36. 'Gates Comments on Chinese J-20,' *Air Force Magazine*, January 10, 2011.
37. Purnell, Danielle, 'Lockheed Martin rolls-out final F-22 Raptor,' *Air Force Reserve Command Official Website*, December 15, 2011. Wolf, Jim, 'U.S. to mothball gear to build top F-22 fighter,' *Reuters*, December 13, 2011. Gertler, Jeremiah, *Air Force F-22 Fighter Program*, Congressional Research Service Report for Congress, July 11, 2013.
38. The Republic of China Armed Forces based on Taiwan remained officially in a state of war with the Chinese mainland.
39. Bumiller, Elisabeth and Wines, Michael, 'Test of Stealth Fighter Clouds Gates Visit to China,' *New York Times*, January 11, 2011.
40. Bumiller, Elisabeth and Wines, Michael, 'Test of Stealth Fighter Clouds Gates Visit to China,' *New York Times*, January 11, 2011. Gates, Robert M., *Duty: Memoirs of a Secretary at War* (New York: Alfred A. Knopf, 2014), Chapter 14: At War to the Last Day.
41. Gates, Robert M., *Duty: Memoirs of a Secretary at War* (New York: Alfred A. Knopf, 2014), Chapter 14: At War to the Last Day.
42. Wines, Michael and Bumiller, Elisabeth, 'Test Unrelated to Gates Visit, China Says,' *New York Times*, January 12, 2011.
43. Levine, Adam, 'Gates: Chinese further along than thought on stealth fighter,' *CNN*, January 10, 2011.
44. 'China's J-20: future rival for air dominance?,' *Strategic Comments*, vol. 17, issue 1, 2011, pp. 1–3.
45. 'China's J-20: future rival for air dominance?,' *Strategic Comments*, vol. 17, issue 1, 2011, pp. 1–3.
46. Donald, David, 'Improved Chinese Stealth Fighter Nears First Flight,' *AIN Online*, February 27, 2014.
47. All prototypes were eventually delivered to the China Flight Test Establishment in Xi'an, although some first spent time at the Flight Test and Training Centre in Cangzhou. '2013' and '2015,' for example, were transferred to Xi'an in April 2015 shortly before the appearances of their successors. New technologies and subsystems were often tested through integration onto the older airframes, a notable example being the re-engining of '2011' to test new powerplants with improved stealth capabilities. Load tests using large drop tanks and testing the firing of air-launched-missiles were reportedly among the other purposes they served.
48. Gertler, Jeremiah, *Air Force F-22 Fighter Program*, Congressional Research Service Report for Congress, July 11, 2013.
49. Thomson, Laura, 'The ten most expensive military aircraft ever built,' *Air Force Technology*, May 30, 2019. Liu, Zhen, 'J-20 vs F-22: how China's Chengdu J-20 "Powerful Dragon" compares with US' Lockheed Martin F-22 Raptor,' *South China Morning Post*, July 28, 2018. 'PLA's J-20 fighters years away from mass production,' *Asia Times*, July 31, 2018.
50. Vartabedian, Ralph and Hennigan, W. J., 'F-22 program produces few planes, soaring costs,' *Los Angeles Times*, June 16, 2013.
51. Grazier, Dan, 'Deceptive Pentagon Math Tries to Obscure $100 Million+ Price Tag for F-35s,' *Project on Government Oversight*, November 1, 2019. As China's first ever stealth fighter, the research and development burden on the J-20 program was considerably higher than for its American counterparts which made its much lower development cost all the more outstanding. While few figures are available for the J-20's R&D costs before service entry all estimates place them at around one tenth those of the F-35.
52. Rogoway, Tyler, 'China's Own "Catfish" Flying Avionics Testbed For The J-20 Fighter Emerges,' *The Drive*, February 3, 2017.
53. Lin, Jeffrey and Singer, Peter W., 'Chinese Stealth Fighter J-20 Starts Production,' *Popular Science*, December 28, 2015.
54. Rupprecht, Andreas, *Modern Chinese Warplanes: Chinese Air Force – Comat Aircraft and Units*, Houston, Harpia, 2018, p. 40. Joe, Rick, 'China's J-20 Stealth Fighter Today and Into the 2020s,' *The Diplomat*, August 16, 2019.
55. Insinna, Valerie, 'Air Force Declares F-35A Ready for Combat,' *Defense News*, August 2, 2016.
56. 'China's first stealth fighter J-20 enters service with Air Force,' *The State Council Information Office of the People's Republic of China Official Website*, March 13, 2017.
57. '新一代隐身战斗机歼-20列装空军作战部队' ('New generation stealth fighter J-20 enters service with Air Force combat force'), *Ministry of National Defence of the People's Republic of China Official Website*, February 9, 2018.
58. Hohmann, James, 'The Daily 202: U.S. came "much closer" to war with North Korea in 2017 than the public knew, Trump told Woodward,' *Washington Post*, September 16, 2020. 'Yes, The United States Did Draw Up A Plan To Drop 80 Nuclear Weapons On North Korea,' *The Drive*, November 27, 2020.
59. Blanchard, Ben and Oliphant, James, 'Chinese State Media Says China Should be Neutral if North Korea Attacks the U.S.,' *Time*, August 11, 2017.
60. Lo, Kinling, 'China "shoots down incoming missiles" during exercise over waters close to North Korea,' *South China Morning Post*, September 5, 2017. Gao, Charolette, 'China's Air Force Tests Missile Defense Near North Korea Border,' *The Diplomat*, September 7, 2017. Wang, Christine, 'China reportedly boosts defense preparations along North Korean border,' *CNBC*, July 24, 2017. Berlinger, Joshua, 'China closes off big chunk of Yellow Sea for military drills,' *CNN*, July 26, 2017. 'How China Has Joined Russia in Drawing a Red Line to Prevent U.S. Military Intervention on the Korean Peninsula,' *Military Watch Magazine*, September 12, 2017.
61. 'China's stealth jet may have done flyover of S Korea,' *Asia Times*, December 7, 2017. 'China Allegedly Sends Cutting-Edge Jet to Eye US-S Korea Joint Air Drill,' *Sputnik News*, December 10, 2017.
62. Global Times on Twitter, 'Former #PLA Air Force commander, General Wang Hai, passed away on Sunday morning in Beijing, aged 94, Chinese media reported. Wang shot down 9 enemy aircraft during the Korean War, with his team engaging the US Air Force on over 80 occasions and downing 29 enemy aircraft.,' August 2, 2020.
63. '震懾台獨分子? 中國最新戰機殲-20部署浙江 15分鐘抵台' ('To deter Taiwan independence elements? China's latest fighter J-20 deployed in Zhejiang and arrived in Taiwan in 15 minutes'), *CNews*, November 28, 2021.
64. Ng, Teddy, 'China deploys J-20 stealth fighter "to keep tabs on Taiwan",' *South China Morning Post*, July 27, 2019.
65. Solen, Derek, 'Second Combat Brigade of PRC Air Force Likely Receives Stealth Fighter,' *China Aerospace Studies Institute*, May 2021.
66. Chan, Minnie, 'China deploys J-20 stealth fighter jets to units monitoring Taiwan Strait,' *South China Morning Post*, June 26, 2021.
67. Liu, Xuanzun, 'J-20 stealth fighter in service for 2nd PLA ace force with home developed engines,' *Global Times*, June 18, 2021.

68. '南部战区第5航空旅开始换装歼-20' ['The 5th Aviation Brigade of the Southern Theater Command Began to Refit with the J-20'], *YouWuQiong*, December 11, 2021.
69. 'J-20A Mighty Dragon production progress,' *Scramble*, December 28, 2023.
70. 'All Chinese Theatre Commands Now Deploy J-20 Fighters: New Stealth Air Brigade Confirmed as Fleet Growth Accelerates,' *Military Watch Magazine*, April 20, 2022.
71. Clark, Colin, 'B-21 Bomber Estimate By CAPE: $511M A Copy,' *Breaking Defense*, September 19, 2016. 'China Building Two New Fields With 230 Nuclear ICBM Silos Able to Strike America – Reports,' *Military Watch Magazine*, July 28, 2021.
72. Insight into the development of the J-20's airframe was provided by a 2001 paper on the aerodynamic configuration of a next generation fighter by Dr. Song Wencong, the mentor of the aircraft's chief designer Yang Wei. The paper highlighted a focus in the industry on resolving design conflicts between stealth, high manoeuvrability, supercruise, transonic performance, and high angle of attack performance. It proposed employment of lift-body leading edge root extension canard configuration which would be unstable in both the lateral and yaw directions. It further proposed all-moving vertical stabilisers significantly smaller than those which would usually be used on a fighter of its size, as well as S-shaped air intakes, in order to improve and reduce the radar cross section. Also proposed were small aspect ratio wings with medium back sweep angle and relatively large dihedral canards. These features, which closely resembled those on the final J-20, were expected to provide excellent stealth capabilities and supersonic drag characteristics, as well as the lift characteristics, stability and controllability needed to achieve high angles of attack.
73. 'Embedded Refueling Probe Seen on China's J20 Stealth Fighter Jet,' *Defense World*, November 15, 2018. 'J-20 now capable of aerial refueling,' *China Military*, November 15, 2018.
74. Li, Jiayao, 'China confirms use of mature stealth testing technology,' *Global Times*, May 22, 2018.
75. Cenciotti, David, 'China's new stealth fighter's missile launch rails prove Beijing can improve U.S. technology,' *The Aviationist*, March 26, 2013.
76. '杨伟代表：歼20将进行系列化发展 不断提升作战能力' ('Representative Yang Wei: The J-20 will undergo serialised development to continuously enhance its combat capabilities.'), *Renmin Wang*, March 20, 2018.
77. Hodge, Nathan, 'China's J-20 Fighter: Stealthy or Just Stealthy-Looking?,' *Wall Street Journal*, January 19, 2011. Interview with Justin Bronk: 'Why the US Military Worries About Chinese Air Power,' *Military Aviation History*, September 9, 2021.
78. Guo, Zhanzhi and Chen, Yingwen and Ma, Lianfeng, '鸭翼的雷达散射截面影响研究 ['Research on the Radar Cross Section Effect of Canards'], *Chengdu Aircraft Design Institute*, September 10, 2019.
79. Bevilaqua, Paul, 'Inventing the F-35 Joint Strike Fighter,' 2009 Wright Brothers Lectureship in Aeronautics, Aerospace Research Central, June 15, 2012 (https://arc.aiaa.org/doi/10.2514/6.2009- 1650). Joe, Rick, 'J-20: The Stealth Fighter That Changed PLA Watching Forever,' *The Diplomat*, January 11, 2021.
80. 'PAK-FA's New Engines Make It "Easily the Best 5th Gen Fighter in the World",' *Sputnik News*, July 25, 2017.
81. Gu, Songfen, 'Strategic Study of China's Fighter Aircraft Development,' *Fighter Aircraft Development Validation Group*, March 2003.
82. Panda, Ankit, 'China's Fifth-Generation Stealth Fighter Is in Combat Service—But With Improved Fourth-Generation Engines,' *The Diplomat*, February 13, 2018. Donald, Davis, 'Paris 2011: Chinese Chengdu J-20 Fighter Still Shrouded in Mystery,' *AIN Online*, June 23, 2011. 'WS-10 engine has 7 variants equipped in 5 fighters including J-20,' *China Arms*, September 11, 2019.
83. 'After the burn,' *Key Aero*, September 10, 2021.
84. Trade Registers, Stockholm International Peace Research Institute Arms Transfer Database (https://armstrade.sipri.org/armstrade/ page/trade_register.php).
85. 'Stealth fighter soon powered by local engines,' *China Daily*, March 13, 2017.
86. 'Does China's J-20 Rival Other Stealth Fighters?,' *China Power: Centre for Strategic and International Studies*, January 2018. Yeung, Jessie and Lendon, Brad, 'China is sending its most advanced fighter jet to patrol disputed seas,' *CNN*, April 15, 2022.
87. Liu, Xuanzun, 'Chinese engine-equipped J-20 fighter proves plateau ability,' *Global Times*, January 19, 2022.
88. Yeung, Jessie and Lendon, Brad, 'China is sending its most advanced fighter jet to patrol disputed seas,' *CNN*, April 15, 2022.
89. Waldron, Greg, 'China's enigmatic J-20 powers up for its second decade,' *Flight Global*, December 28, 2020.
90. 'PAK-FA's New Engines Make It "Easily the Best 5th Gen Fighter in the World",' *Sputnik News*, July 25, 2017.
91. Yeo, Mike and Pocock, Chris, 'More J-20 Stealth Fighters Built in China,' *AIN Online*, July 19, 2016. Roblin, Sebastien, 'Is China About to Give Its Best Fighters Powerful New Jet Engines?,' *National Interest*, March 5, 2020.
92. 'J-20 stealth fighter's capabilities to be enhanced,' *China Daily*, March 13, 2018.
93. 'J-20 Begins Flying With Groundbreaking WS-15 Engine: Will China's Stealth Fighter be the World Leader in Thrust?,' *Military Watch Magazine*, January 15, 2022.
94. Liu, Xuanzun, 'China's WS-15 turbofan engine undergoes tests, shows improved performance,' *Global Times*, March 14, 2022.
95. 'China: J-20's WS-15 engine undergoes multiple rounds of testing,' *China Arms*, March 15, 2022.
96. Liu, Xuanzun, 'China fully capable of jet engine development, narrowing technology gaps,' *Global Times*, March 9, 2022.
97. Liu, Xuanzun, 'China's WS-15 turbofan engine undergoes tests, shows improved performance,' *Global Times*, March 14, 2022.
98. Chan, Minnie, 'Is China's W-15 engine to power J-20 stealth fighter jet nearing completion?,' *South China Morning Post*, December 19, 2022.
99. Waldron, Greg, 'Chinese executive hints at progress with J-20's new WS-15 engine,' *Flight Global*, March 27, 2023.
100. Chen, Chuanren, 'China May Have Flown J-20 With Domestic WS-15 Engines,' *Aviation Week*, June 30, 2023.
101. Liu, Xuanzun, 'China's WS-15 turbofan engine undergoes tests, shows improved performance,' *Global Times*, March 14, 2022.
102. Episkopos, Mark, 'Out of Range: Why China's J-20 Might Have the Tools to Kill an F-35,' *National Interest*, June 20, 2019. '霹雳15在200公里外直取预警机，歼20的优势太大了' ('The Thunderbolt 15 took the early warning aircraft 200 kilometres away, and the advantage of the J-20 is too great'), *Sina*, January 16, 2021. '让美国顾忌的歼20"御用武器"竟然也要出口了' ('The J-20's "Crown Weapon" that Raises American Doubts is About To Be Exported'), *Xinhua*, September 28, 2021.
103. International Institute for Strategic Studies, *The Military Balance*, Volume 119, 2019, p. 8.
104. Lee, Tae-Woo, *Military Technologies of the World: Volume 1 and 2* (Westport: Praeger, 2009), p. 82. Trimble, Stephen, 'New long-range missile project emerges in US budget,' *Flight Global*, November 2, 2017. Chopra, Anil, 'Air Defence of India: Evolving Options,' *Indian Defence Review*, vol. 34.3, July-September 2019.
105. International Institute for Strategic Studies, *The Military Balance*, Volume 119, 2019, p. 237.
106. International Institute for Strategic Studies, *The Military Balance*, Volume 119, 2019, p. 237. Fisher, Jr., Richard D., *China's Military Modernization: Building for Regional and Global Reach* (Westport: Praeger, 2008), p. 232. Chan, DM, 'China's military can't underestimate AIM-260,' *Asia Times*, June 27, 2019.

107. Drew, James, 'USAF seeks "interim" CHAMP, longer-range air-to-air missiles,' *Flight Global,* September 17, 2015. 'Military Strategy Forum: General Herbert "Hawk" Carlisle on Air Combat Command: Today's Conflicts and Tomorrow's Threats,' *Center for Strategic & International Studies,* September 18, 2015.
108. 'Lockheed Developing AIM-260 To Counter China's PL-15,' *Aviation Week,* June 26, 2019. Barrie, Douglas, 'AIM-260 missile: the US Air Force and beyond-visual-range lethality,' *International Institute for Strategic Studies Military Balance Blog*, October 24, 2019.
109. Newdick, Thomas, 'A Guide To China's Increasingly Impressive Air-To-Air Missile Inventory,' *The Drive*, September 1, 2022.
110. International Institute for Strategic Studies, *The Military Balance*, Volume 119, 2019, p. 237.
111. Newdick, Thomas, 'A Guide To China's Increasingly Impressive Air-To-Air Missile Inventory,' *The Drive*, September 1, 2022.
112. Bronk, Justin, 'Russian and Chinese Combat Air Trends: Current Capabilities and Future Threat Outlook,' Royal United Services Institute, October 2020.
113. Rupprecht, Andreas, *Modern Chinese Warplanes: Chinese Air Force – Comat Aircraft and Units* (Houston: Harpia, 2018), p. 37.

Chapter 3

1. 'Stealth Fighter or Bomber?,' *The Diplomat*, July 26, 2011.
2. Waldron, Greg, 'China's enigmatic J-20 powers up for its second decade,' *Flight Global*, December 28, 2020.
3. Axe, David, 'China's Over-Hyped Stealth Jet,' *The Diplomat*, January 7, 2011. Kopp, Carlo and Goon, Peter, 'Chengdu J-XX [J-20] Stealth Fighter Prototype: A Preliminary Assessment,' *AusAirpower, Technical Report APA-TR-2011-0101*, January 3, 2011. 'J-20 Design,' *Global Security* (https://www.globalsecurity.org/military/world/china/j-20-design.htm).
4. 'Stealth Fighter or Bomber?,' *The Diplomat*, July 26, 2011.
5. Joe, Rick, 'J-20: The Stealth Fighter That Changed PLA Watching Forever,' *The Diplomat*, January 11, 2021.
6. 'J-20 stealth fighter's capabilities to be enhanced,' *China Daily*, March 13, 2018.
7. '杨伟代表：我的一个梦想，未来战机由中国来制定标准' ('Representative Yang Wei: One of my dreams is that China will set the standards for future fighter jets'), State-Owned Assets Supervision and Administration Commission of the State Council, March 20, 2018.
8. 'Embedded Refueling Probe Seen on China's J20 Stealth Fighter Jet,' *Defense World*, November 15, 2018.
9. Waldron, Greg, 'China's enigmatic J-20 powers up for its second decade,' *Flight Global*, December 28, 2020.
10. Lei, Zhao, 'J-20 fighter takes part in first combat exercises,' *China Daily*, January 12, 2018.
11. 'Air Force reveals J-20 combat formation,' *People's Daily*, January 21, 2020.
12. Liu, Xuanzun and Guo, Yuandan, 'China, Russia joint drills conclude with live-fire anti-terrorism operation featuring J-20,' *Global Times*, August 13, 2021.
13. '英媒：中国将研发新版歼-20隐身战机 准备启动6代机项目' ('British media: China will develop a new version of the J-20 stealth fighter and prepare to launch the 6th generation program'), *Sina,* March 15, 2018.
14. Gouré, Daniel, 'The Air Force Needs A New High-Low Mix,' *Lexington Institute,* June 9, 2016. Isachenkov, Vladimir, 'Russia, China sign roadmap for closer military cooperation,' *AP News,* November 23, 2021.
15. Hollings, Alex, 'Why the F-117 Nighthawk Is Such a Badass Plane,' *Popular Mechanics*, August 12, 2019. 'Celebrating a quarter century of Stealth,' *Flight Global*, June 19, 2007.
16. Liu, Xuanzun, 'J-20 fighters conduct nocturnal battle drill to hone stealth advantages,' *Global Times,* January 16, 2022.
17. '新一代隐身战斗机歼-20列装空军作战部队' ('New generation stealth fighter J-20 enters service with Air Force combat force'), *Ministry of National Defence of the People's Republic of China Official Website*, February 9, 2018.
18. 'China Tests its Newest and Most Dangerous Fighters: J-20 and J-16 Further Integrated into the Air Force,' *Military Watch Magazine*, January 12, 2018. Dominguez, Gabriel, 'China's J-20 fighter aircraft takes part in its first combat exercise, says report,' *Jane's 360*, January 15, 2018. '祖孙隔代青春对话：歼-20 飞行员陈浏军礼致敬姥爷刘玉堤' ('Youth dialogue between grandparents and grandchildren: J-20 pilot Chen Liujun pays tribute to his grandfather Liu Yudi'), *Ministry of Defence of the People's Republic of China Official Website*, May 28, 2021.
19. '新一代隐身战斗机歼-20列装空军作战部队' ('New generation stealth fighter J-20 enters service with Air Force combat force'), *Ministry of National Defence of the People's Republic of China Official Website*, February 9, 2018.
20. Lo, Kinling, 'New fighter jets will be used in "more regular" patrols over South China Sea,' *South China Morning Post,* February 13, 2018.
21. 'China's J-20 Fifth-Gen Fighter Successfully Completes Combat Training at Sea,' *Sputnik News*, May 9, 2018.
22. 'Experts: J-20 will surely patrol Taiwan in future,' *China Military Online*, May 18, 2018.
23. 'China Deploys YJ-12B and HQ-9B Missiles on South China Sea Islands,' *Navy Recognition*, May 4, 2018.
24. Minnick, Wendell, 'China Expands Presence With Fighters on Woody Island,' *Defence News*, November 8, 2015. 'China deploys PLANAF J-11BH/BHS fighters to Woody Island,' *Alert 5*, October 31, 2015.
25. 'China Is Threatening to Fly Its J-20 Stealth Fighter over Taiwan,' *National Interest,* May 21, 2018.
26. 'Taipei Threatened by China's New Stealth Fighters? Why the J-20 is Unlikely to See Action over the Taiwan Strait,' *Military Watch Magazine*, May 29, 2018. 'Russia Now Has "Treasure Trove" of Info About Stealthy F-22s – US General,' *Sputnik News*, January 5, 2018.
27. 'J-20 fighter takes part in nighttime exercise in air superiority role,' *People's Daily,* June 2, 2018.
28. Rogoway, Tyler, 'China's J-20 Stealth Fighter Stuns By Brandishing Full Load Of Missiles At Zhuhai Air Show,' *The Drive*, November 11, 2018.
29. Liu, Zhen, 'China-India border dispute: advanced fighter jets sent to nearby airbases,' *South China Morning Post*, August 19, 2020.
30. '媒体披露歼20战机真实空战实力 以"零损伤"击落敌机17架' ('The media disclosed the real air combat strength of the J-20 fighter jets and shot down 17 enemy planes with "zero damage"'), *163*, September 25, 2020. 'Chinese Media Claims J-20 Shot Down 17 "Enemy" Fighters in Exercises – How Reliable Are These Claims?,' *Military Watch Magazine*, September 16, 2020.
31. Pickrell, Ryan, 'Air Force F-35s Wrecked Enemies During Mock Air Combat,' *Task and Purpose*, February 21, 2019.
32. Liu, Xuanzun, 'China's J-20 stealth fighter jet flies without Luneburg lens, shows combat readiness,' *Global Times*, April 5, 2021.
33. Luneburg reflective lenses are designed to increase the aircraft's radar signature and thus mask its true stealth capabilities.
34. Liu, Xuanzun, 'China's J-20 stealth fighter jet flies without Luneburg lens, shows combat readiness,' *Global Times*, April 5, 2021.
35. Liu, Xuanzun, 'J-20 shows high combat readiness in New Year combat training,' *Global Times*, January 9, 2022. Liu, Xuanzun, 'Newly commissioned J-20 stealth fighters on combat alert,' *Global Times*, January 17, 2022.
36. Liu, Xuanzun, 'J-20 fighters conduct nocturnal battle drill to hone stealth advantages,' *Global Times,* January 16, 2022.
37. *Battle Stations*, Season 1, Episode 39, F-22 Raptor.
38. 'Aerospace Nation: Gen Kenneth S. Wilsbach,' *Mitchell Institute for Aerospace Studies* (YouTube Channel), March 15, 2022.

39. 'Aerospace Nation: Gen Kenneth S. Wilsbach,' *Mitchell Institute for Aerospace Studies* (YouTube Channel), March 15, 2022.
40. Liu, Xuanzun, 'J-20 fighter jet starts routine training patrols in East, South China Seas,' *Global Times*, April 13, 2022.
41. Yeung, Jessie and Lendon, Brad, 'China is sending its most advanced fighter jet to patrol disputed seas,' *CNN*, April 15, 2022.
42. 'J-20 has ability to defend nation's airspace,' *China Daily*, August 15, 2022.
43. 'J-20 fighters to escort Y-20 aircraft for 1st time in repatriating remains of CPV martyrs from S.Korea,' *Global Times*, September 14, 2022.
44. '2架殲-20戰機為接運英雄的運-20護航'('Two J-20 fighter jets escorted the Y-20 transporting the military heroes') *People's Daily*, November 24, 2023.
45. Liu, Xuanzun, 'China's J-20 stealth fighter jets drive away foreign aircraft in combat patrols over East China Sea,' *Global Times*, October 13, 2022.
46. Liu, Zhen, 'China outlines J-20 stealth fighter's role in intercepting foreign warplanes by releasing footage that may show rare encounter with F-35,' *South China Morning Post*, January 18, 2023.
47. Liu, Xuanzun, 'PLA Air Force's Wang Hai Air Group fully equipped with J-20 fighter jets, expels foreign aircraft by giving full play to stealth capability,' *Global Times*, January 17, 2023.
48. Huang, Kristin, 'Chinese bombers, fighter jets in new year military drills appear to have boosted combat capacity,' *South China Morning Post*, January 15, 2023.
49. Liu, Xuanzun and Guo, Yuandan, 'Chinese Air Force unit that emerged victorious from Korean War vows to be ready for combat,' *Global Times*, July 27, 2023.
50. 'China's J-20 fighter jet, YY-20 refueling plane amaze audience at Changchun Airshow,' *CGTN*, July 27, 2023. 'Highlights of Changchun Air Show,' *Xinhua*, July 27, 2023.
51. Gady, Franz-Stefan, 'China's First 5th Generation Fighter Moves Into Serial Production,' *The Diplomat*, October 31, 2017. Johnson, Reuben F, 'China's J-20 fifth-gen fighter moves into series production,' *Jane's*, October 26, 2017.
52. 'Chinese J-20 Fighter, Y-20 Transport Aircraft Designs Finalized, Ready for Mass Production: Expert,' *Defense World*, November 12, 2017. Tirpak, John A., 'China Likely Stepping Up Stealth Fighter Production,' *Air and Space Forces*, October 8, 2021. Chan, Minnie, 'World-class production lines speed up deliveries of China's J-20 stealth jet fighter,' *South China Morning Post*, November 27, 2022.
53. Liu, Xuanzun, 'China ramps up J-20 stealth fighter production after domestic engine switch,' *Global Times*, December 12, 2021.
54. Liu, Xuanzun, 'China ramps up J-20 stealth fighter production after domestic engine switch,' *Global Times*, December 12, 2021.
55. Liu, Xuanzun, 'China ramps up J-20 stealth fighter production after domestic engine switch,' *Global Times*, December 12, 2021.
56. Tirpak, John A., 'China Likely Stepping Up Stealth Fighter Production,' *Air Force Magazine*, October 8, 2021.
57. 'Top designers announce J-20 production capacity, Y-20 domestic engines, new carrier-based fighter at Airshow China,' *Global Times*, September 29, 2021.
58. Liu, Xuanzun, 'Newly commissioned J-20 stealth fighters on combat alert,' *Global Times*, January 17, 2022.
59. 'China to speed up research into new strategic weapons for air force: J-20 chief designer,' *Global Times*, September 27, 2021.
60. 'U.S. Sources Estimate China Likely Already Fields Over 200 J-20 Stealth Fighters: Just How Many Are There?,' *Military Watch Magazine*, April 17, 2022.
61. Liu, Xuanzun, 'PLA to train pilots faster with new programs amid warplane production capacity boost,' *Global Times*, July 24, 2022.
62. Liu, Xuanzun, 'China ramps up J-20 stealth fighter production after domestic engine switch,' *Global Times*, December 12, 2021.
63. 'Technical Challenges Delay Mass Production of F-35 Stealth Fighter Yet Again,' *Military Watch Magazine*, January 3, 2021. 'America's F-35 Stealth Fighter Delayed From Full Scale Production Again,' *Military Watch Magazine*, October 30, 2020. 'Full Scale F-35 Production Further Delayed: Fighter Suffers More Breakdowns Than Expected – Pentagon,' *Military Watch Magazine*, November 15, 2019.
64. Gertler, Jeremiah, *Air Force F-22 Fighter Program*, Congressional Research Service Report for Congress, July 11, 2013. Battle Stations, Season 1, Episode 39, F-22 Raptor.
65. Joe, Rick, 'China's J-20 Gets Another Upgrade,' *The Diplomat*, August 1, 2023.
66. Andreas Rupprecht on Twitter: 'The J-20 production rate ! I have to admit, although I've now accepted a much higher production rate of the J-20, this sounds almost too good to be true: "120 J-20 production per year. at least 500 j-20s by 2025" Both Shilao and Ayi have claimed J-20 production was in 3 digit.', July 18, 2023.

Chapter 4

1. Looking to foreign examples, the F-22's very high operational costs led the U.S. Air Force to continue to comprise the bulk of its fleet of lighter F-16s and later F-35s. By contrast the relatively low difference in operational costs between Russian Flankers and lighter MiG-29s when compared to the discrepancy in performance led the Russian Air Force to favour a slightly smaller fleet of more heavyweight Flankers as the more cost-effective option.
2. Gordon, Yefim and Davidson, Peter, *Sukhoi Su-27 Flanker* (North Branch: Specialty Press, 2006), pp. 41, 64.
3. 'Aerospace Nation: Gen Kenneth S. Wilsbach,' *Mitchell Institute for Aerospace Studies* (YouTube Channel), March 15, 2022.
4. Tirpak, John A., 'Divestitures and Purchases: USAF's 2023 Aircraft Plans,' *Air Force Magazine*, April 29, 2022.
5. Sixth generation fighters were referred to by prominent figures in the U.S. aerospace sector as necessary to be able to 'go one-v-five, one-v-ten, etc.' against fifth generation fleets, meaning they were intended to have overwhelming performance advantages over fighters like the J-20 where F-22s and F-35s could only 'go head-to-head from a one-v-one perspective.' Whether with upgrades including integration of sixth generation technologies the J-20 could avoid such a disadvantage was potentially decisive in how long it could remain in production. ('A next-gen digital backbone will give U.S. platforms a one-versus-many advantage against near-peers,' *Breaking Defense*, October 12, 2022.)
6. Fisher Jr, Richard D., China's Global Military Power Projection Challenge to the United States, Testimony Before the House Permanent Select Committee on Intelligence, United States House of Representatives, May 17, 2018.
7. Andreas Rupprecht on Twitter: 'The J-20 production rate ! I have to admit, although I've now accepted a much higher production rate of the J-20, this sounds almost too good to be true: "120 J-20 production per year. at least 500 j-20s by 2025" Both Shilao and Ayi have claimed J-20 production was in 3 digit.', July 18, 2023.
8. Andreas Rupprecht on Twitter: 'The J-20 production rate ! I have to admit, although I've now accepted a much higher production rate of the J-20, this sounds almost too good to be true: "120 J-20 production per year. at least 500 j-20s by 2025" Both Shilao and Ayi have claimed J-20 production was in 3 digit.', July 18, 2023.
9. Gertler, Jeremiah, *Air Force F-22 Fighter Program*, Congressional Research Service Report for Congress, July 11, 2013.
10. Losey, Stephen, 'F-35 delivery delays to cost Lockheed hundreds of millions in 2023,' *Defense News*, July 20, 2023. Gomes, Nathan, 'Lockheed Martin trims F-35 jet delivery outlook after supplier delays,' *Reuters*, September 7, 2023.
11. Losey, Stephen, 'First upgraded F-35s won't be ready for combat until next year,' *Defense News*, April 24, 2024.
12. Liu, Xuanzun, 'Chinese engine-equipped J-20 fighter proves plateau ability,' *Global Times*, January 19, 2022.

13. Liu, Xuanzun, 'Innovation of twin-seat J-20 stealth fighter to lead world, military experts say after reported maiden flight,' *Global Times*, November 7, 2021.
14. Liu, Xuanzun, 'Innovation of twin-seat J-20 stealth fighter to lead world, military experts say after reported maiden flight,' *Global Times*, November 7, 2021. Tate, Andrew, 'China may be developing first two-seat stealth combat aircraft,' *Jane's*, January 18, 2019. 'J-16 vs. J-10C: Chinese Pilot Reveals Which Elite Fighter is Superior,' *Military Watch Magazine,* March 26, 2021.
15. Huang, Kristin, 'China's air force unveils new twin-seat J-20 stealth fighter jet in video,' *South China Morning Post*, January 12, 2021.
16. *Ordinance Industry Science Technology*, August 2022.
17. Gordon, Yefim and Davidson, Peter, *Sukhoi Su-27 Flanker* (North Branch: Specialty Press, 2006), Chapter 4.
18. Chan, Minnie, 'China's stealth fighter goes into mass production after thrust upgrade,' *South China Morning Post*, July 12, 2020.
19. A minority of analysts believed that the J-20B designation was in fact that designation of the twin seater, and that new variant with WS-15 engines was designated J-20A while single seaters produced from 2015–2019 and 2019–2023 with either the AL-31 or the WS-10C were all baseline 'J-20' models.
20. 'J-20 stealth fighter's capabilities to be enhanced,' *China Daily*, March 13, 2018.
21. 'J-20 stealth fighter's capabilities to be enhanced,' *China Daily*, March 13, 2018.
22. Xuanzun, Liu, 'J-20 fighter could get directed-energy weapon, drone-control capability: experts,' *Global Times*, January 23, 2022.
23. Gordon, Yefim, *Sukhoi Su-27* (Hinckley: Midland Publishing, 2007), p. 182.
24. Heath, Timothy R. and Gunness, Kristen and Cooper III, Cortez A., 'The PLA and China's Rejuvenation: National Security and Military Strategies, Deterrence Concepts, and Combat Capabilities,' (Santa Monica: RAND, 2016), pp. 38–39.
25. Chan, DM, 'Stealth wars: China's J-20 vs USAF's F-35,' *Asia Times*, July 30, 2019.
26. *China Military Power: Modernising a Force to Fight and Win*, Washington DC, Defense Intelligence Agency, 2019, p. 85.
27. Roblin, Sebastien, 'Meet The FB-22 Stealth Bomber: It Would Have Been Russia's Worst Nightmare,' *1945*, March 29, 2022.
28. Sweetman, Bill, 'Smarter Bomber,' *Popular Science,* June 25, 2002. Tirpak, John, 'Long Arm of the Air Force,' *Air Force Magazine*, October 2002. Cortes, Lorenzo, 'Roche Looking to Next Year for Near-Term Proposals on Strike Concepts,' *Defense Daily,* March 18, 2004. Vago Muradian, 'F-22 May be Modified as Speedy New Medium Bomber to Strike Moving Targets,' *Defense Daily International,* January 18, 2002.
29. Gordon, Yefim and Davidson, Peter, *Sukhoi Su-27 Flanker* (North Branch: Specialty Press, 2006), pp. 75, 76, 80, 81. 'Meet the Su-34, Russia's Supersonic Strike Aircraft NATO Fears,' *National Interest,* June 4, 2018. Су-34 (Su-34), *Sukhoi Official Website* (https://www. sukhoi.org/products/samolety/254/).
30. Huang, Kristin, 'Why two heads would be better than one for China's "Mighty Dragon" fighter jet,' *South China Morning Post,* April 27, 2021.
31. Xuanzun, Liu, 'J-20 fighter could get directed-energy weapon, drone-control capability: experts,' *Global Times*, January 23, 2022.
32. Chan, Minnie, 'Drones to become "loyal wingmen" for China's advanced J-20 stealth fighter jets, state media reports,' *South China Morning Post*, October 20, 2022.
33. Chan, Minnie, 'China's J-20 stealth fighter joins the People's Liberation Army air force,' *South China Morning Post*, March 10, 2017.
34. *Military and Security Developments Involving the People's Republic of China 2020*, Annual Report to Congress, United States Department of Defence, pp. 51, 56.
35. Rogoway, Tyler, 'China's J-20 Stealth Fighter Stuns By Brandishing Full Load Of Missiles At Zhuhai Air Show,' *The Drive*, November 11, 2018.
36. Cone, Allen, 'Lockheed's Sidekick adds increased firepower to F-35 fighters,' *Lockheed Martin*, May 3, 2019.
37. *Military and Security Developments Involving the People's Republic of China 2021*, Annual Report to Congress, United States Department of Defence, p. 144.
38. Liu, Xuanzun, 'Military developing airborne laser attack pod, says report,' *Global Times*, January 7, 2020.
39. Xuanzun, Liu, 'J-20 fighter could get directed-energy weapon, drone-control capability: experts,' *Global Times*, January 23, 2022.
40. Liu, Xuanzun, 'J-20 fighter pilot calls for improved military communications network development,' *Global Times*, March 9, 2023.
41. 'Mitchell Institute's Deptula on China's J-20 Stealth Fighter,' *Defense and Aerospace Report* (YouTube Channel), November 11, 2016.
42. Liu, Xuanzun, 'J-20 fighter pilot calls for improved military communications network development,' *Global Times*, March 9, 2023.
43. Liu, Xuanzun, 'Test pilot sees China's J-20 to get 2D thrust vectoring nozzles,' *Global Times*, April 19, 2021.
44. 开了眼了on Twitter: 'Let's enjoy it together. (with video attachment),' August 27, 2022 (https://twitter.com/RupprechtDeino/status/1563407148260151297).
45. 乐子壬1号机 on Twitter: 'WS-15? (with video attachment),' September 20, 2022 (https://mobile.twitter. com/kt396/ status/1572175436783976450?fbclid=IwAR2dJMySeZzz NSLJVFaKZPaHGlu7fhEN3077AemtHmXPn_ PzeuGBiITIxHw&fs=e&s=cl).
46. '在见到J20之前我没想过一架飞机还能这样飞' ('before seeing the J-20 I never conceived that an aircraft could fly like this'), 大卫坑 (Billibilli Account), November 14, 2022.
47. Koryakin, Oleg, 'Первый и последний: чем был уникален истребитель Су-37' ('First and last: what made the Su-37 fighter unique'), *Rossiyskaya Gazeta,* April 2, 2020.
48. Chen, Stephen, 'Chinese scientists hail "incredible" stealth breakthrough that may blind military radar systems,' *South China Morning Post*, July 19, 2019.
49. Chen, Stephen, 'Chinese scientists create a "plasma shower" to improve stealth bomber performance,' *South China Morning Post*, August 26, 2022. Gordon, Yefim, *Sukhoi S-37 and Mikoyan MFI: Russian Fifth-Generation Fighter Technology Demonstrators* (Hinckley: Midland Publishing, 2001), pp. 21–22.
50. Chen, Stephen, 'China's heat-seeking radar with 300km range boosts anti-stealth tech, say defence scientists,' *South China Morning Post*, August 23, 2022.
51. Freedberg Jr., Sydney J., 'F-35 Ready For Missile Defense By 2025: MDA Chief,' *Breaking Defense,* April 11, 2018.
52. LaGrone, Sam, 'Video: Successful F-35, SM-6 Live Fire Test Points to Expansion in Networked Naval Warfare,' *United States Naval Institute*, September 13, 2016. Abbott, Rich, 'New Demonstration Shows F-35's Data Sharing Capability,' *Aviation Today,* August 8, 2019. Osborn, Kris, 'The F-35 Strengthens Its Role in Missile Defense,' *National Interest,* November 30, 2021.
53. Rogoway, Tyler, 'The Airborne Laser May Rise Again But It Will Look Very Different,' *Jalopnik*, August 18, 2015.
54. Trevithick, Joseph and Rogoway, Tyler, 'F-22 And F-35 Datalinks Finally Talk Freely With Each Other Thanks To A U-2 Flying Translator,' *The Drive*, April 30, 2021. Insinna, Valerie, 'Here's why the Valkyrie drone couldn't translate between F-35 and F-22 jets during a recent test,' *C4ISRN*, December 19, 2020. Tirpak, John A., 'Skyborg Drone Translates Between F-35 and F-22 in Test,' *Air Force Magazine,* December 16, 2020.
55. 'Mitchell Institute's Deptula on China's J-20 Stealth Fighter,' *Defense and Aerospace Report* (YouTube Channel), November 11, 2016.

56. Harper, Jon, 'Kendall: Air Force's next-gen fighter will be "better than anything" China can produce,' *Defense Scoop*, September 29, 2022.
57. 'Department of Defense Press Briefing by Secretary James and Gen. Goldfein on the State of the Air Force in the Pentagon Briefing Room,' *Department of Defense Official Website*, August 10, 2016.
58. 'A Twenty-First-Century Sputnik Moment: China's Mozi Satellite,' *Nippon.com,* August 13, 2019. Kwon, Karen, 'China Reaches New Milestone in Space-Based Quantum Communications,' *Scientific American,* June 25, 2020. Chen, Stephen, 'China uses quantum satellite to protect world's largest power grid against attacks,' *South China Morning Post*, December 10, 2021. 'China Builds the World's First Integrated Quantum Communication Network,' *Scitech Daily,* January 6, 2021. Chen, Yu-Ao et al., 'An integrated space-to-ground quantum communication network over 4,600 kilometres,' *Nature,* vol. 589, January 2021, pp. 214–219.
59. International Institute for Strategic Studies, *The Military Balance*, Volume 120, 2020, p. 236. Rogoway, Tyler, 'China's Type 055 Super Destroyer Is A Reality Check For The US And Its Allies,' *The Drive,* June 28, 2017.
60. Rupprecht, Andreas and Newdick, Thomas, 'China's Massive Fleet Of Radar Planes And The Strategy Behind It,' *The Drive*, April 5, 2023.
61. Lin, Jeffrey and Singer, P. W., 'The World's Largest Military Plane in Production is China's Y-20,' *Popular Science*, June 20, 2016.
62. 'Aerospace Nation: Gen Kenneth S. Wilsbach,' *Mitchell Institute for Aerospace Studies* (YouTube Channel), March 15, 2022.
63. 'Aerospace Nation: Gen Kenneth S. Wilsbach,' *Mitchell Institute for Aerospace Studies* (YouTube Channel), March 15, 2022.
64. Liu, Xuanzun, 'China's J-20 stealth fighter jets drive away foreign aircraft in combat patrols over East China Sea,' *Global Times*, October 13, 2022.
65. Grier, Peter, 'A Quarter Century of AWACS,' *Air Force Magazine*, March 1, 2002. Veronico, Nic and Dunn, Jim, *21st Century U.S. Air Power,* Grand Rapids, MI, Zenith Imprint, 2004, p. 83. LaFayette, Ken, 'The E-3 Sentry Airborne Warning and Control System,' *Warfare History Network*.
66. Li, Junsheng and Chen, Bo and Hou, Na, *Cooperation for a Peaceful and Sustainable World: Part 2* (Bingley: Emerald, 2013), pp. 18–19.
67. Yeo, Mike, 'China ramps up production of new airborne early warning aircraft,' *Defense News*, February 5, 2018.
68. 'Aerospace Nation: Gen Kenneth S. Wilsbach,' *Mitchell Institute for Aerospace Studies* (YouTube Channel), March 15, 2022. Tirpak, John A., 'USAF Selects Boeing's E-7A Wedgetail as Successor to AWACS,' *Air & Space Forces*, February 28, 2023.
69. International Institute for Strategic Studies, *The Military Balance*, Volume 122, 2022, Chapter Three: North America, and Chapter Six: Asia.
70. Rupprecht, Andreas, 'Images confirm Y-20U aerial tanker is in PLAAF service,' *Jane's*, November 30, 2021.
71. Liu, Xuanzun, 'China's J-16D electronic warfare aircraft starts combat training, "to team up with J-20 stealth fighter",' *Global Times*, November 6, 2021.
72. 'China Modifies H-6G Bomber into Electronic Warfare Aircraft,' *Defense Mirror,* January 22, 2018. Withington, Thomas, 'Jamming JDAM: The Threat to US Munitions from Russian Electronic Warfare,' *Royal United Services Institute*, June 6, 2023.
73. Erickson, Andrew S. and Lu, Hanlu and Bryan, Kathryn and Septembre, Samuel, 'Research, Development, and Acquisition in China's Aviation Industry: The J-10 Fighter and Pterodactyl UAV,' *Study of Innovation and Technology in China: University of California Institute of Global Conflict and Cooperation*, no. 8, 2014.
74. Hao, Karen, 'Yes, China is probably outspending the US in AI—but not on defense,' *MIT Technology Review,* December 5, 2019. Li, Daitian and Wang, Tony W. and Xiao, Yangao, 'Is China Emerging as the Global Leader in AI?,' *Harvard Business Review,* February 18, 2021. Schmidt, Eric et al., *Final Report*, National Security Commission on Artificial Intelligence, March 2021 (https://www.nscai.gov/wp-content/uploads/2021/03/Full-Report-Digital-1.pdf). Vincent, James, 'China and the US are battling to become the world's first AI superpower,' *The Verge,* August 3, 2017. Kia, On Wong Wilson, 'China's AI Strike Force on COVID-19,' *Asian Education and Development Studies,* vol. 10, no. 2, 2021, pp. 250–262.
75. Kress, Bob and Gillcrist, Paul, 'Battle of the Superfighters: F-14D vs. F/A-18E/F (Two Experts Say the Super Hornet Isn't so Super),' *Flight Journal*, February 2002. Thompson, Roger, *Lessons Not Learned; The U.S. Navy's Status Quo Culture*, Annapolis, Naval Institute Press, 2007, Chapter 10.
76. 'China decided to compare Su-35 and J-10C fighters in training aerial combat,' *Top War*, June 16, 2020.
77. 'J-10C beats J-16 and Su-35, winning the most "Golden Helmets",' *China Arms*, December 15, 2021.
78. Interview with Justin Bronk: 'Why the US Military Worries About Chinese Air Power,' *Military Aviation History*, September 9, 2021.
79. The value of Chinese 'tanker killer' capabilities was notably highlighted by a 2008 RAND Corporation study, which found that to sustain operations for 130 combat coded F-22s from Guam over Taiwan the USAF would need to launch three to four tanker sorties every hour to deliver 2.6 million gallons of fuel. Even assuming no F-22 combat losses and a perfect combat performance, the projected loss of tankers would leave the American jets unable to return to their bases and result in extreme losses among the Raptors. Guam was at the time seen as the nearest location where bases had a chance of surviving PLA strikes. By limiting tanker operations over very wide areas, PL-XX armed J-16s could thus provide a very effective complement to J-20 units over the East and South China Seas. Stillion, John and Perdue, Scott, 'Air Combat Past, Present and Future,' Project Air Force briefing, August 2008, Unclassified/FOUO/Sensitive, Slide 29. Watts, Barry, 'The F-22 Program in Retrospect,' Center for Strategic and Budgetary Assessments, August 2009.
80. *Military and Security Developments Involving the People's Republic of China 2020*, Annual Report to Congress, United States Department of Defense, p. 76.

Chapter 5

1. 'Aerospace Nation: Gen Kenneth S. Wilsbach,' *Mitchell Institute for Aerospace Studies* (YouTube Channel), March 15, 2022.
2. *Military and Security Developments Involving the People's Republic of China 2020*, Annual Report to Congress, United States Department of Defense, 2020, p. 75.
3. Finnerty, Ryan, 'Overtaxed F-35 engines rack up $38 billion in extra maintenance costs,' *Flight Global,* June 2, 2023. Trevithick, Joseph, 'F-35 Engine Running Too Hot Due To "Under-Speccing," Upgrade Now Vital,' *The Drive*, March 30, 2023. Cohen, Rachel S., 'Software glitch during turbulence caused Air Force F-35 crash in Utah,' *Air Force Times*, July 28, 2023. 'F-35's Troubled F135 Engine is Causing Unavailability Rates at 600% of Standard Levels,' *Military Watch Magazine*, September 11, 2022.
4. 'Only 30% of American F-35s Are Fully Mission Capable: Program Chief Slams "Unacceptable" Fleet Performance,' *Military Watch Magazine*, April 3, 2023.
5. Guertin, Nickolas H., 2022 Office of the Director, Operational Test and Evaluation Annual Report, January 2023, p. 62. Reim, Garrett, 'Lockheed Martin F-35 has 873 deficiencies,' *Flight Global,* January 31, 2020. Grazier, Dan, 'F-35 Design Flaws Mounting, New Document Shows,' *Project on Government Oversight,* March 11, 2020. Capaccio, Anthony, 'Lockheed's F-35 Quality Flaws Persist as Production Ramps Up,' *Bloomberg*, October 16, 2023.

6. 'Thunder Without Lightning, The High Costs and Limited Benefits of the F-35 Program,' *National Security Network*, August 2015.
7. Axe, David, 'Pentagon's big budget F-35 fighter "can't turn, can't climb, can't run",' *Reuters*, July 14, 2014.
8. Grazier, Dan, 'F-35 Continues to Stumble,' *Project on Government Oversight*, March 30, 2017.
9. Roy, Ananya, 'F-35 jets have over 200 deficiencies, unlikely to be combat ready by 2018-2019, Pentagon report says,' *International Business Times,* January 17, 2017. 'F-35 "scarcely" fit to fly: Pentagon's chief tester,' *Press TV*, April 3, 2017.
10. Grazier, Dan, 'F-35 Far from Ready to Face Current or Future Threats, Testing Data Shows,' *Project on Government Oversight*, March 19, 2019.
11. Britzky, Haley, 'Acting SecDef Shanahan thinks the F-35 program is "f–ked up" just like everyone else,' *Task and Purpose,* April 25, 2019.
12. McCain, John, 'U.S. Senator, Arizona, Opening statement by SASC Chairman John McCain on the F-35 Joint Strike Fighter Program,' April 26, 2016. Parsons, Dan, 'Blistering Highlights From The Latest F-35 Sustainment Hearing,' *The Drive*, May 9, 2022.
13. Britzky, Haley, 'Acting SecDef Shanahan thinks the F-35 program is "f–ked up" just like everyone else,' *Task and Purpose,* April 25, 2019.
14. Tirpak, John A., 'F-35 Production Challenged to Keep Up with Demand,' *Air & Space Forces*, July 3, 2023.
15. 'Why the F-35's Low Altitude Ceiling is a Major Drawback: Adversaries Will Almost Always Strike From Above,' *Military Watch Magazine*, August 4, 2023.
16. Mehta, Aaron, 'Boeing Positions F-15 as F-22 Supplement,' *Breaking Defense*, September 15, 2015.
17. Cenciotti, David, '"If we don't keep F-22 Raptor viable, the F-35 fleet will be irrelevant" Air Combat Command says,' *The Aviationist*, February 4, 2014.
18. 'China's stealth jet may be ready this year, US commander says,' *Straits Times*, May 2, 2018.
19. Johnson, Reuben F., 'Myths Of The Raptor,' *CBS News*, June 28, 2009.
20. Grazier, Dan, '108 U.S. F-35s Won't Be Combat-Capable,' *National Interest,* October 16, 2017. Trevithick, Joseph, 'USMC's Older F-35Bs May Only Be Able To Fly Around A Quarter Of Their Expected Service Life (Updated),' *The Drive*, February 1, 2019.
21. Chan, Minnie, 'Is China ready for aircraft carrier No 4? Talk swirls over stealth fighter jets at PLA naval base,' *South China Morning Post*, May 1, 2022.
22. Rizzo, Jennifer and Keyes, Charley, 'Is China closer than thought to matching U.S. fighter jet prowess?,' *CNN*, January 6, 2011.
23. Cordesman, Anthony H. and Colley, Steve and Wang, Michael, *Chinese Strategy and Military Modernization in 2015: A Comparative Analysis* (Washington DC: Centre for Strategic and International Studies, 2015), p. 285.
24. 2014 Report to Congress of the U.S.-China Economic and Security Review Commission, One Hundred Thirteenth Congress, Second Session, November 2014, p. 311.
25. Dutton, Peter and Erickson Andrew S. and Martinson, Ryan, *China's Near Seas Combat Capabilities* (Newport: Naval War College, China Maritime Studies Institute, 2014).
26. Lendon, Brad, 'China's new J-20 stealth fighter screams on to scene,' *CNN*, November 1, 2016.
27. Snow, Shawn, 'Don't expect the Corps' new Chinese J-20 stealth fighter to be dogfighting with Marine jets,' *Marine Corps Times*, December 21, 2018.
28. Demerly, Tom, 'USAF Confirms: The Chinese J-20 Spotted In Georgia Is a Mock-Up Used For Training by the U.S. Marine Corps,' *The Aviationist*, December 9, 2018.
29. Malyasov, Dylan, 'US Marines train with fake Chinese J-20 fighter jet,' *Defence Blog*, October 21, 2023.
30. Audit of Training Ranges Supporting Aviation Units in the U.S. Indo-Pacific Command, Department of Defense, Office of the Inspector General, April 17, 2019.
31. Trevithick, Joseph, 'F-22 Raptors Could Become Aggressors Under Air Force's Radical New Training Plan,' *The Drive*, June 23, 2020.
32. Rogoway, Tyler, 'F-117 Nighthawks Now Appear To Be Flying As Adversaries In Red Flag Aerial War Games,' *The Drive*, November 27, 2020.
33. Vanover, Christie, 'Nellis AFB aggressors, F-35 pilots "punish" blue air to develop unstoppable force,' *United States Air Force Official Website*, August 4, 2021. Newdick, Thomas, 'F-35s Have Flown Their First "Red Air" Missions As Dedicated Stealth Aggressor,' *The Drive*, August 4, 2021.
34. Pickrell, Ryan, 'A US F-22 Raptor pilot describes the challenge of going up against F-35 red air aggressors,' *Business Insider,* August 30, 2021.
35. Hollings, Alex, 'Stealth Death Match: China's J-20 Vs. F-22 Raptor (Who Wins?),' *1945*, April 24, 2022.
36. Trevithick, Joseph, 'Stealthy Target Drones Sought As QF-16 Program Winds Down,' *The Drive*, August 23, 2022.
37. Tirpak, John A., 'New Study: USAF Needs Big Cash Infusion to Overcome Aging Fighter Fleet,' *Air & Space Forces Magazine*, June 29, 2023. Guastella, Joseph and Birkey, Douglas and Gunzinger, Eric and Poling, Aidan, 'Accelerating 5th Generation Airpower: Bringing Capability and Capacity to the Merge,' *Mitchell Institute*, vol. 43, June 29, 2023.
38. Tirpak, John A., 'New Study: USAF Needs Big Cash Infusion to Overcome Aging Fighter Fleet,' *Air & Space Forces Magazine*, June 29, 2023. Guastella, Joseph and Birkey, Douglas and Gunzinger, Eric and Poling, Aidan, 'Accelerating 5th Generation Airpower: Bringing Capability and Capacity to the Merge,' *Mitchell Institute*, vol. 43, June 29, 2023.
39. 'Only 30% of American F-35s Are Fully Mission Capable: Program Chief Slams "Unacceptable" Fleet Performance,' *Military Watch Magazine*, April 3, 2023. Abrams, A. B., 'South Korean Defense Sources Express Concerns About Unreliable F-35 Fighters,' *The Diplomat*, October 7, 2022. Freedberg Jr., Sydney J., 'F-35 readiness and flight hours fell in 2022, says CBO,' *Breaking Defense*, February 13, 2023.
40. Tirpak, John A., 'New Study: USAF Needs Big Cash Infusion to Overcome Aging Fighter Fleet,' *Air & Space Forces Magazine*, June 29, 2023. Guastella, Joseph and Birkey, Douglas and Gunzinger, Eric and Poling, Aidan, 'Accelerating 5th Generation Airpower: Bringing Capability and Capacity to the Merge,' *Mitchell Institute*, vol. 43, June 29, 2023.
41. LaGrone, Sam, 'CNO Greenert: Navy's Next Fighter Might Not Need Stealth, High Speed,' *United States Naval Institute*, February 4, 2015.
42. Tirpak, John A., 'Raptor 01,' *Air Force Magazine*, July 1997, p. 48.
43. Davies, Steve and Dildy, Doug, *F-15 Eagle Engaged, The World's Most Successful Jet Fighter* (Oxford: Osprey, 2007), p. 12. Rininger, Tyson V., *F-15 Eagle At War* (Minneapolis: Zenith Press, 2009), p.16. Jenkins, Dennis R., *McDonnell Douglas F-15 Eagle: Supreme Heavy-Weight Fighter* (Leicester: Midland Publishing, 1998), pp. 7–8. Gordon, Yefim, *Mikoyan MiG-25 Foxbat: Guardian of the Soviet Borders* (Hersham: Ian Allen Publishing, 2007), pp. 98–101.
44. Davies, Peter E. and Thornborough, Tony, *McDonell Douglas F-15 Eagle* (Ramsbury: Crowood Press, 2001), p. 8.
45. Rogoway, Tyler, 'The Air Force's Secret Next Gen Air Dominance Demonstrator Isn't What You Think It Is,' *The Drive*, September 21, 2020.
46. Trevithick, Joseph, 'Air Force Generals Aren't "Losing Sleep" Over China's J-20 Stealth Fighter,' *The Drive*, September 23, 2022.
47. Tirpak, John A., 'Piecing Together the NGAD Puzzle,' *Air and Space Forces*, April 29, 2022.

48. Trevithick, Joseph, 'Air Force Generals Aren't "Losing Sleep" Over China's J-20 Stealth Fighter,' *The Drive*, September 23, 2022.
49. 'NGAD: USAF's sixth generation fighter is on schedule, acquisition official says,' *Aerospace Manufacturing*, October 11, 2021.
50. Tirpak, John A., 'Kelly Worries F-35 Flying Costs Won't Hit Target, and That China May Get NGAD First,' *Air Force Magazine*, February 26, 2021.
51. 'NGAD: USAF's sixth generation fighter is on schedule, acquisition official says,' *Aerospace Manufacturing*, October 11, 2021.
52. Altman, Howard, 'Air Force's Next Generation Air Dominance "Fighter" Program Enters New Stage,' *The Drive*, June 2, 2022.
53. Insinna, Valerie and Losey, Stephen, 'US Air Force bails on Mattis-era fighter jet readiness goal,' *Defense News*, May 8, 2020.
54. 'Here's How the F/A-18 Super Hornet Will Be Able to Fight Against J-20 and Su-57 Stealth Fighters,' *The Aviation Geek Club*, October 26, 2017.
55. Helfrich, Emma, 'F-35 Will Get New Radar Under Massive Upgrade Initiative,' *The Drive*, January 3, 2023.
56. Newdick, Thomas and Rogoway, Tyler, 'The F-22 Raptor Could Finally Get The Infrared Sensor It Was Originally Promised,' *The Drive*, January 13, 2022.
57. Pickrell, Ryan, 'A US F-22 Raptor pilot describes the challenge of going up against F-35 red air aggressors,' *Business Insider*, August 30, 2021.
58. 'Aerospace Nation: Gen Kenneth S. Wilsbach,' *Mitchell Institute for Aerospace Studies* (YouTube Channel), March 15, 2022. Waldron, Greg, 'E-3 insufficient for timely detection of J-20: Pacific Air Forces chief,' *Flight Global*, March 17, 2022.
59. Tirpak, John A., 'USAF Selects Boeing's E-7A Wedgetail as Successor to AWACS,' *Air & Space Forces*, February 28, 2023.
60. Axe, David, 'New Chinese Missile has USAF Spooked,' *Real Clear Defense*, September 24, 2015.
61. Trimble, Steve, 'The Weekly Debrief: Does Raytheon's New AIM- 120D3 Beat China's Best Missile?,' *Aviation Week*, July 25, 2022. Tirpak, John A., 'Piecing Together the NGAD Puzzle,' *Air and Space Forces*, April 29, 2022.
62. Trimble, Steve, 'The Weekly Debrief: Does Raytheon's New AIM-120D3 Beat China's Best Missile?,' *Aviation Week*, July 25, 2022. 'US Air Force, Raytheon Missiles & Defense execute first live-fire test of AMRAAM F3R,' *Raytheon Missiles and Defense*, July 18, 2022.
63. Department of the Air Force Presentation to the House Armed Services Committee Subcommittee on Air and Land Forces, United States House of Representatives, Subject: Air Force Programs, Combined Statement of: Lieutenant General Daniel J. Darnell, Air Force Deputy Chief Of Staff For Air, Space and Information Operations, Plans And Requirements (AF/A3/5), Lieutenant General Mark D. Shackelford, Military Deputy, Office of the Assistant Secretary of the Air Force for Acquisition (SAF/ AQ), [and] Lieutenant General Raymond E. Johns, Jr., Air Force Deputy Chief of Staff for Strategic Plans And Programs (AF/A8), May 20, 2009, pp. 7, 8.
64. Tirpak, John A., 'Raptor 01,' *Air Force Magazine*, July 1997, p. 48.
65. Axe, David, 'China's Over-Hyped Stealth Jet,' *The Diplomat*, January 7, 2011.
66. Allison, Graham, 'America second? Yes, and China's lead is only growing,' *Boston Globe*, May 22, 2017.
67. McGill, Earl J., *Black Tuesday Over Namsi: B-29s vs MiGs – The Forgotten Air Battle of the Korean War* (Solihull: Helion & Company, 2012), Chapter 4: The Machinery of War.
68. Stone, I. F., *Hidden History of the Korean War* (Boston: Little, Brown and Company, 1988), p. 342.
69. Stone, I. F., *Hidden History of the Korean War* (Boston: Little, Brown and Company, 1988), p. 342.

ABOUT THE AUTHOR

Abraham Abrams is an expert in East Asian military affairs and security. He holds a masters degree in War Studies from King's College London with a dissertation focusing on the post-Cold War modernisation of the Chinese People's Liberation Army, as well as a prior Masters in East Asian International Relations, and is currently pursuing a related PhD. Having followed the J-20 program's development closely since before its entry into service, the writer is a long-time analyst of China's defence sector and major weapons programs as well as broader trends in international security affecting China and its neighbours. His assessment of the fighter program thus places it in historical and geopolitical contexts to provide a fuller understanding of its significance.